The Fire and EMS Department Safety Officer

GORDON M. SACHS

Prentice Hall
Upper Saddle River, NJ 07458

Library of Congress Cataloging-in-Publication Data

Sachs, Gordon M.

The fire and EMS department safety officer / Gordon M. Sachs.

p. cm.

Includes index.

ISBN 0-13-013917-3

1. Fire extinction—Safety measures. 2. Emergency management—Safety measures. 3. Firefighters. 4. Safety engineers. I. Title.

TH9182.S23 2000

363.11′962892—dc21

00-055760

Publishing Unit President: Gary Lee June
Acquisitions Editor: Katrin Beacom
Production Editor: BookMasters, Inc.
Production Liaison: Larry Hayden IV
Director of Manufacturing and Production: Bruce Johnson
Managing Editor: Patrick Walsh
Manufacturing Buyer: Ilene Sanford
Art Director: Marianne Frasco
Marketing Manager: Kristin Walton
Editorial Assistant: Douglas Stokes
Cover Design: Maria Guglielmo-Walsh
Composition: BookMasters, Inc.
Printing and Binding: R.R. Donnelley & Sons, Harrisonburg, Virginia

Prentice-Hall International (UK) Limited, *London*
Prentice-Hall of Australia Pty. Limited, *Sydney*
Prentice-Hall Canada Inc., *Toronto*
Prentice-Hall Hispanoamericana, S. A., *Mexico*
Prentice-Hall of India Private Limited, *New Delhi*
Prentice-Hall of Japan, Inc., *Tokyo*
Prentice-Hall Singapore Pte. Ltd.
Editora Prentice-Hall do Brasil, Ltda., *Rio de Janeiro*

10 9 8 7 6 5 4 3 2 1
ISBN 0-13-013917-3

Contents

Preface

There is nothing more important than the life of an emergency responder. This person, dedicated to protecting the lives of strangers by putting themselves in harm's way, plays a critical role in society. Unfortunately, this role is often summed up after a tragedy, such as that which occurred on December 4, 1999, when six firefighters were killed in a fire in Worcester, Massachusetts:

> "Today the sun didn't rise. It didn't rise because last night we lost six members of our families. They were our brothers, our fathers, our sons, and our friends. And they were our protectors. Yes, they were firefighters, but more importantly they were members of our family and today our entire city grieves."
>
> —Worcester Mayor Ray Mariano, the morning after the tragedy

> ". . . They were firefighters to the core, heroes already, as we have heard, to their friends and loved ones, not to mention the people they saved through the years. For all six, being a firefighter was more than a job, it was in their blood. So when they went into that building that night, they were following their dream to serve, to save lives, and to stick together.
>
> "Like their fellow firefighters everywhere, they embodied the best of our nation—of commitment and community, of teamwork and trust—values at the core of our character; values reflected in the daily service not only of those we lost, but in this awesome parade of men and women who have come from all over our country and from some countries beyond our borders to honor their comrades and console their families.
>
> "Too often, we take them for granted, our firefighters. In the days ahead, I hope every American will find an occasion to thank those in their communities who stand ready every day to put their lives on the line when the alarm bell rings."
>
> —President Bill Clinton, at the memorial service in Worcester

How do we protect the emergency responder—firefighters and emergency medical personnel—while they face the risks associated with their jobs? By providing them with proper equipment, giving them proper training, developing comprehensive standard operating procedures or guidelines, and instilling a positive, safety-oriented attitude. Still, sometimes that's not enough.

Fire and EMS departments need to focus on safety and health at the station, in the apparatus, and at the incident scene. To do this, an individual—the safety officer—must be dedicated to the administrative and incident-related aspects of the emergency services. They must be an advocate of the responder, someone whose mission is to help keep the responders alive.

This book provides guidance for the individual serving in the role of fire or EMS department safety officer. This individual may be the department's health and safety officer, whose primary role is to look at safety from a program standpoint to prevent injuries, illnesses, exposures, and accidents involving emergency responders. Or, it might be a line officer appointed at an emergency scene to be the incident safety officer. This officer must understand and recognize safety cues and have the fortitude to identify unsafe conditions, operations, or acts and to take the appropriate measures to prevent harm to the operating crews.

This book is divided into three sections, plus a comprehensive set of appendices. Part 1 addresses general concepts and requirements, including an overview of the position of safety officer and information on laws, regulations, and standards affecting the position. The topic of record keeping and documentation is addressed in this section, as well. Part 2 addresses administrative aspects, keying on the role of the health and safety officer. The key topics in this section include risk management, health maintenance, and accident and injury investigation. Part 3 addresses operational aspects, including emergency incident risk management and operating at emergency incidents. Obviously, this section is important for anyone who may serve as incident safety officer. The appendices contain information and material that will be useful to a fire or EMS department safety officer. In addition to points-of-contact and synopses of laws, regulations, and standards that will be useful to any health and safety officer, there are forms and checklists for use at emergency incidents of all types by the incident safety officer.

It is important to note that the fire and EMS department safety officer is but one link—albeit a very important link—in the chain of safety. Each responder needs to look out for his or her own safety. Each officer must look out for the safety of their crew. Each battalion or division chief or division, group, or sector officer must look out for the safety of everyone assigned to or operating within their assigned area. Each fire or EMS chief or incident commander must look out for the safety of everyone in the department or operating at the incident. The safety officer, to put it succinctly, is an important part of this last link—the chief's risk manager, or the eyes and ears of the incident commander—in the chain that keeps emergency responders safe and healthy.

Acknowledgments

This book is dedicated to the men and women of America's emergency services who make it their goal to help keep other responders safe. Specifically recognized are health and safety officers of fire and EMS departments, infection control officers of fire and EMS departments, and chief officers who have made responder safety and health a high priority within their departments. Of special importance are those early fire department safety officers—the "safety zealots," who built safety and health programs before there were regulations or standards requiring departments to develop them and who had little guidance on how to build a program or be a safety officer. Having held such a position, I hold my former colleagues from departments across the country in high esteem.

This book is also dedicated to members of the National Fire Protection Association (NFPA) technical committees, whose hard work and perseverance resulted in standards that have significantly impacted the safety of today's responder. Of special importance are the current and former members of the NFPA Technical Committee on Fire Service Occupational Safety and Health and the NFPA Technical Committee on Fire Service Occupational Health and Medical. These individuals worked hard—not for personal gain, but to help make the emergency responder safer and keep them healthy.

Further, this book is also dedicated to the course developers and instructors of the courses in the National Fire Academy's Health and Safety curriculum. These courses—from the FSCO and FHSP courses of the mid- and late 1980s to the ICERP, ISO, and HSO courses today—have had a tremendous impact on the number of firefighters killed and injured in the line of duty. These course developers and instructors have helped put safety on the front burner in today's fire and emergency medical services by serving as leaders, mentors, and role models as they practice what they preach.

The preceding individuals and teams have provided the catalyst for the transition to a "safety culture" in the American emergency services. The numbers bear this out: overall, the number of firefighters killed in the line of duty has dropped dramatically since the mid-1980s; NFA's first specific firefighter safety course was released in 1986, and NFPA 1500 was first published in 1987.

I would be remiss if I did not include a dedication to my family. They mean everything to me, and they've put up with a lot over the years. Writing this book wasn't all that bad, but the years of working in the emergency services, especially in the safety and health arena, took its toll. The stress, the overtime, the meetings at all hours, the pager . . . if you are a health and safety officer, you understand what I mean. Their understanding and support has been a godsend for me.

My wife Lisa, who typed most of the manuscript for this book, is a firefighter and EMS responder, as well as the public educator for our local fire and EMS department. My son Adam is also a firefighter and EMT and has entered the U. S. Navy as a hospital corpsman to stay in the medical field. My daughter Brandy has joined the fire and EMS department; her goal is to be a confined-space rescue technician. My guess is that they all got involved in the emergency services so we'd be able to spend more time as a family (albeit at the station or on a call). I am proud to say that the "safety bug" has rubbed off onto each of them. Simply, I am proud of them, and am proud to be a part of their family.

A special thanks goes to the reviewers listed below for their feedback and suggestions:

Chief W. David Bunce
Rural/Metro Fire Department
Scottsdale, AZ

Raymond E. Hawkins
VFIS, Director of Education and Training
York, PA

Steven I. Weissman
Safety Officer, Fairfax County
Virginia Fire and Rescue Department
Fairfax, VA

NOTE: All references to NFPA standards in this book relate to the current edition of the standard as of the time the text was written.

- NFPA 1500 (1997 edition), Standard on Fire Department Occupational Safety and Health Program
- NFPA 1521 (1997 edition), Standard on Fire Department Safety Officer
- NFPA 1561 (1995 edition), Standard on Fire Department Incident Management System
- NFPA 1581 (2000 edition), Standard on Fire Department Infection Control Program
- NFPA 1582 (2000 edition), Standard on Medical Requirements for Fire Fighters and Information for Fire Department Physicians

About the Author

Gordon M. Sachs

Gordon M. Sachs has over 23 years of fire service and EMS experience, including over 10 years as a chief officer in both career and volunteer departments. Currently, he is chief of Fairfield Fire and EMS in Pennsylvania and is the director of the IOCAD Emergency Services Group, a consulting firm providing management and technical support and solutions for the emergency services across the country. He holds a Master of Public Administration Degree, a Bachelor of Science degree in education, and is a graduate of the National Fire Academy's Executive Fire Officer program. Chief Sachs is the author of *Officer's Guide to Fire Service EMS* (Fire Engineering Books, 1999) and has coauthored several books and publications, including *Fire Department Occupational Health and Safety Handbook* (NFPA, 1998) and *The Fire Chief's Handbook, Fifth Edition* (Fire Engineering Books, 1995). He has also has written numerous trade journal articles and made several major presentations in the areas of fire service/EMS leadership and management, emergency responder health and safety, and incident command. He is a course developer and adjunct instructor at the National Fire Academy and is a former U.S. Fire Administration Program Manager. He has been a member of several national committees, including some NFPA Technical Committees on safety and health issues, as well as various federal and state committees on emergency service issues.

Part One

General Concepts and Requirements

1

Introduction

In the late 1980s, the fire service began to champion the position of safety officer. Today, most fire departments have someone whose responsibilities include staying abreast of safety issues and concerns. In EMS, this concept often fails to go beyond a designated infection control officer. Many EMS organizations now also recognize the wisdom of expanding the scope of "EMS Safety" to include more than infection control. Just staying abreast of the federal and local mandates to promote worker safety alone can be overwhelming. Add the many other aspects of safety that should be addressed in the emergency services, and it is easy to see why the need for a designated safety officer exists. Even though safety is everyone's responsibility, the safety officer holds the key to reducing injuries and deaths in both the fire service and EMS workplaces. Thousands—even hundreds of thousands—of dollars stand to be saved through the effective implementation of an EMS safety officer.

The Safety Officer

Little guidance is available on the EMS-related responsibilities of the safety officer. Fire service models for basic safety standards and the role of the safety officer can be found in the 1997 editions of NFPA 1500, *Fire Department Occupational Safety and Health Program,* and NFPA 1521, *Fire Department Safety Officer.* These can be applied to EMS organizations.

According to these standards, a person in the role of safety officer is responsible for knowing the following.

- Current local, state, and federal laws pertaining to occupational safety and health that apply to fire service and EMS professionals
- Factors related to physical and mental fitness and basic health, such as basic knowledge of exercise physiology, good nutrition, and stress management
- How to properly and effectively manage a program designated for safety and health

The safety officer has a job that requires appropriate experience with and knowledge of emergency operations. It demands a certain degree of literacy, the ability to work with data for record-keeping purposes, skills in personnel management, and a knack for interagency relations.

A safety officer can be part of any size or type of fire or EMS organization. Small, volunteer, or rural services may rotate the administrative safety officer position and designate someone to be the safety officer on a per-scene basis. Medium-sized organizations may delegate safety officer functions to the training officer, because there are many crossover tasks. A large organization can benefit tremendously by having a full-time safety officer.

An important element of the safety officer's role is to be able to report directly to whoever is the highest-ranking leader of the organization. With so much responsibility for the well-being of personnel, the safety officer's ability to do the job must not be jeopardized by others along the chain of command; this could compromise the mission if health or safety issues arise in the areas supervised by resentful or suspicious intermediate managers.

Emergency Services Safety Considerations

The roles and tasks of the safety officer consist primarily of incident operations and administrative functions. Only the environment, dynamics, and pace change; the safety officer's mission remains essentially the same. Because it is impossible for one person to be present at each fire or EMS call even in the smallest organizations, one ongoing educational task for any safety officer is to permeate crews with a focus on safety.

The roles and tasks of the safety officer include the following.

- Respond—or send a trained designee—to major incidents or those that involve (or may involve) unusual safety hazards. At the scene, the safety officer must not engage in patient care or triage activities but instead remain dedicated to the safety function. This person's job is to notice (and mitigate, where possible) hazardous or careless actions and to assess for changing conditions or circumstances that could increase the existing hazard level. The safety officer should report to the incident commander on arrival at the site. These two people should work together closely on behalf of the field providers and the patients.
- Assess personnel at incident scenes—especially those of prolonged duration—to determine who needs on-site rehabilitation. Even temporary relief of duties allows for rest and rehydration. The safety officer can arrange for a rehab area away from the action and for an EMS crew (and the local critical incident stress management team, if needed) to staff it.
- Attend post-incident analyses, adding positive and negative reviews of the safety angle, where appropriate.
- Identify and correct safety hazards in the workplace. This includes hazards in the organization's offices, training sites, stations, vehicle repair shops,

supply rooms, and so on. The safety officer is responsible not only for field providers but for support staff and office workers, as well.

- Know the applicable laws, regulations, and requirements as written by federal and state government bodies (such as the Occupational Safety and Health Administration, or OSHA) and national standards-making groups (such as ASTM and the National Fire Protection Association, or NFPA). Routinely assess departmental policies, procedures, and operations for compliance. When compliance is not achieved, members of the organization need to be educated about the risks: There may be large fines or other penalties. Worse, someone could be seriously injured or killed.
- Investigate all on-the-job injuries, illnesses, and exposures (both hazardous materials and infectious diseases). Document the event and take (or recommend) appropriate follow-up action to minimize the chances that similar events could happen again.
- Keep accurate personnel records, including a confidential health database that documents medical examinations, occupational injuries, illnesses, exposures, and medical testing and treatment. These records are vital to support workers' compensation claims. In addition, the safety officer should have records of physical fitness training requirements and achievements.
- Maintain inspection and service records on equipment and facilities used by the organization, including items such as seldom-used personal protective equipment and monitor-defibrillators. The safety officer should know what records the vehicle maintenance department is keeping and whether those records are thorough and accurate.
- Investigate emergency vehicle crashes and keep records of follow-up information. Substance abuse testing sometimes may be needed. The safety officer must know the correct procedures to follow.
- Keep records of situations that pose potential liability to the organization. These situations may include threats to sue, actual lawsuits, and other service complaints.
- Act as the safety liaison to other agencies. In addition, the safety officer should be the liaison with equipment manufacturers (including vehicles), standard-setting authorities (such as protocol committees), and other regulatory agencies (such as OSHA and in-state equivalents). By doing so, the safety officer can maintain a global view of the safety issues faced by the organization and make appropriate recommendations to the safety committee or the organization head.
- Act as a two-way information resource. Listen to suggestions and recommendations of employees and introduce them to management. Relay new information and programs from managers or outside regulatory agencies to field personnel.

The Incident Safety Officer

The safety of responders must be addressed at every emergency. Standard operating procedures (SOPs) address safety at routine operations. At an incident that involves multiple emergency units, the person charged with monitoring the safety of responders at the scene of an emergency is known as the incident safety officer (ISO). The ISO may be the health and safety officer (HSO), or the functions of the ISO may be performed by any member of the fire department designated by the incident commander (IC). At the scene of an emergency incident, the ISO reports directly to the IC.

The ISO is a member of the fire department or agency with knowledge of safety, hazards, and the procedures established within the agency. The ISO does not need to be an officer within the fire department. NFPA 1521 contains requirements for needed knowledge and the job requirements of the ISO. The ISO may be asked to perform some nonemergency follow-up activities, but the majority of the work of the ISO is performed at the scene of an emergency.

The scene of an emergency is one of the most hazardous working areas for a responder. During the effort to control the emergency, and after the incident has been controlled, responders and others on the scene may concentrate more on the work at hand than on the safety of those working at the incident. The only job of the ISO is to focus on the safety of emergency operations and provide the IC with needed safety information and recommendations. The ISO has the authority to stop all operations if they judge that the operations create an imminent hazard to responders.

The Health and Safety Officer

The administrative functions of the safety officer are performed by an individual that more and more departments are calling a health and safety officer (HSO). The HSO is a department-level, primarily administrative position that encompasses responsibility for coordinating the safety and wellness aspects of organizational activities. The HSO guides department policy as it applies to occupational safety practices and member welfare issues. The HSO is responsible for interpreting rules, regulations, and standards and their applications to safety and health issues in the emergency services. Examples include bloodborne and airborne pathogen regulations; hazardous materials training requirements; confined-space rescue training requirements; personal protective equipment; and civil unrest and violent situation protocols. NFPA 1521 also contains requirements for needed knowledge and the job requirements of the HSO.

The HSO must have appropriate functional authority. It does no good to create the position of safety officer (or HSO) or to expand the role to include both fire and EMS without empowering this individual to make difficult, and sometimes unpopular, changes on behalf of safety. Without top-level administrative support (both verbal and financial), the task sometimes may be unreasonably troublesome or even impossible.

SUMMARY

The basic duty of the safety officer (ISO and HSO) is to protect the safety of responders. This individual is, in effect, a risk manager, consultant, adviser, and leader and must address EMS issues as well as fire issues. In most fire departments, EMS accounts for about 70 percent of the response volume.

The safety officer essentially has two separate functions: that of the ISO and that of the HSO. Typically, the ISO is another trained officer appointed at specific incidents; the HSO is primarily an administrative position. Regardless of the approach taken, the bottom line is that safety is as important to EMS providers as it is to firefighters—something that seems to have been forgotten along the way.

2
The Safety Officer

The Incident Safety Officer (ISO)

By the nature of the duties they perform, emergency responders are at risk of death, injury, or illness. Incident safety should be a primary concern of all those who respond to the aid of the community or jurisdiction they serve. To help minimize the risk to responders, one of the ways the incident command system (ICS) provides for responder safety is by giving the incident commander (IC) the ability to appoint and use a safety officer. This position is part of the ICS organization's command staff. Although the IC has overall responsibility for the safety of the responders, the incident safety officer (ISO) has the direct responsibility of focusing on the safety aspects of the incident.

An ISO is an officer at the scene of an incident or training evolution who is responsible for monitoring and assessing safety hazards or unsafe situations and developing measures for ensuring personnel safety (see Figure 2-1).

The National Fire Protection Association (NFPA) has developed a national consensus standard outlining the duties and responsibilities of the ISO. NFPA 1521, *Standard for Fire Department Safety Officer,* covers health, safety, and wellness program management duties as well as the responsibilities of the ISO during an emergency incident.

Both the NFPA standard and the ICS safety officer description give the ISO the authority to alter, suspend, or terminate unsafe acts or hazardous activities. This makes the ISO position unique within the ICS organization. Although the ICS typically follows the chain of command, the ISO can bypass the system to correct unsafe actions or remove responders from the threat of immediate danger. For example, an ISO can remove firefighters from the interior of a structure that has the potential of imminent collapse. An ISO can also remove responders from the area of an overturned vehicle that has not been properly shored to prevent it from rolling over onto them. When the ISO takes action to remove responders from the threat of danger, each immediate supervisor and the IC should be advised as to what action was taken and why the ISO made the determination.

In addition to correcting unsafe acts and hazardous activities, the ISO is responsible for identifying existing or potential hazards that do not present an imminent threat to responder safety. Communicating these hazards to the IC will

FIGURE 2-1 The incident safety officer (ISO) is responsible for identifying hazards that pose an imminent or potential threat to the safety of responders. Photo by Gerry Suftko, Mesa (AZ) Fire Department.

allow the action plan to address the hazards and will help the IC anticipate modifications that may need to be made to the plan.

Basic Duties and Responsibilities of the ISO

The primary responsibility of the ISO at every incident is to protect the safety of the responders. The ISO should ensure that responders follow safe practices. The ISO can do this in several ways. The ISO should see that responders

- wear full personal protective equipment,
- work in teams in hazardous areas,

- have back-up personnel available to react to an unexpected event quickly,
- use an accountability system to track personnel,
- follow departmental and recognized safety practices, and
- follow safe practices during training exercises.

Knowledge of Typical Response Incidents

It is important for the ISO to have the requisite knowledge to function effectively at an incident. The ISO must understand the hazards inherent at a typical incident response. This knowledge must include the following.

Structural fire

- building construction
- fire behavior
- flame spread
- limits to how long firefighters can operate with self-contained breathing apparatus (SCBA)

EMS response

- infection control procedures
- scene security measures
- wearing of personal protective equipment
- critical incident stress indicators and management

Special operations

- safety lines staffed at a water rescue
- proper shoring at a ditch cave-in
- approved lifelines at a high-angle rescue
- product identification at a hazmat incident
- use of technical experts

Without the requisite knowledge, the ISO could endanger the safety of personnel through the inability to recognize when the responders might be at risk.

Work with Departmental Health and Safety Officer (HSO)

If a department has an occupational health and safety program, a designated health safety officer (HSO) manages that program. The HSO develops and coordinates the program for the department. (The HSO is discussed later in this chapter.)

Although the HSO in some departments may also serve as an ISO during an emergency, the primary duties of each position are very different. If these positions are filled by separate individuals, they must work together to improve safety (see Figure 2-2). During an incident, the ISO may observe some practices that need

FIGURE 2-2 The ISO and HSO need to work together to identify and implement long-term changes that will affect responder safety. Photo by Ed Roberts, Dothan (AL) Fire Department.

to be modified in the current standard operating procedure (SOP) or see procedures that need to be developed. Working together, the two can make the necessary changes.

Another advantage the ISO will gain by working with the HSO is to develop a better understanding as to why SOPs were developed. Having this understanding will enable the ISO to identify unsafe practices or to recognize when responders may need treatment or rehabilitation.

Should a responder be injured or killed at an incident, the ISO and HSO will need to work closely to determine the circumstances surrounding the death or injury. They also will need to determine the cause of any illness a responder may contract as a result of response.

ISO as a Risk Manager

The primary responsibility of an ISO is to prevent injuries to the responders. By halting unsafe acts and removing members from threatening environments, the ISO is acting as a risk manager.

To be an effective risk manager, the ISO should have a plan to follow in monitoring conditions and actions at the incident. Much like the IC's action plan, the safety action plan is in the ISO's head at most smaller incidents. At large-scale or long-running incidents, the ISO's safety plan will be an integral part of a written incident action plan.

Making sure a rehabilitation area is established is an important function of the ISO. The ISO should monitor responders to anticipate and recognize when rehabilitation is needed.

As the incident risk manager, the ISO must properly document the safety aspects of the emergency. Good documentation is important for the postincident analysis in the event of an injury or death of a responder and as a part of the department's records for the incident. The postincident analysis should concentrate on positive reinforcement and learning from mistakes. It should not be a punitive exercise.

Characteristics of an Effective ISO

Personal Characteristics

An ISO should possess certain personal characteristics in order to handle the position. Effective communications skills are key. The ISO must be an active listener, demonstrate the ability to understand what others are saying, and know how their messages may relate to incident safety. The ISO must also be able to convey messages in a clear and concise manner. The information the ISO is likely to impart can be critical to protecting responder safety and should be delivered in a manner others can understand clearly.

There are times when the ISO may need to deliver a message the responders do not want to hear. Telling firefighters to get out of a building or a paramedic to glove up may not be well received. Firefighters may feel that they can get the fire out in "just a couple more minutes." The paramedic may say, "Let me get this line started first." The ability to convey messages in a positive manner that convinces the responders that their safety is at issue may require above-average communication skills.

Another personal characteristic of an effective ISO is to have a genuine concern for the safety of those operating at the incident. The ISO needs to understand that his/her role is to protect the safety of the responders rather than to be an enforcer. Although the ISO may be enforcing safe practices and procedures, the reason for doing so is to ensure responder safety; the ISO's job is not to act as the incident bully.

The ISO must be able to focus on safety issues. The ISO may want to become involved in the tactical operations in addition to fulfilling the functions of the ISO. Because safety is such a high priority at an incident, the ISO needs to concentrate his/her full attention on safety and leave the operations to the other responders.

The ISO may need to make quick decisions during times of high stress. An understanding of the incident and the operations being carried out will help the ISO make good decisions. The ability to make those decisions in a decisive and confident manner is a characteristic the ISO must have in order to react quickly enough to keep the responders out of harm's way.

Knowledge of Duties

If operating within a departmental or agency ICS organization, the ISO must have a clear understanding of how the position fits in the incident organization. The ISO must know and understand how to function as part of the IC's command staff; with whom to communicate and coordinate; and what specific job responsibilities are described in the agency's ICS.

A thorough knowledge of the department's or agency's SOPs and any applicable laws, standards, or regulations should also be part of the ISO's background. Through knowledge of SOPs, and as a result of training on tactical operations, the ISO should be able to recognize when responders are using safe operational practices and procedures and when they are not.

DOCUMENTATION REQUIREMENTS

Documentation of the safety aspects of the incident is something the department or agency may require as part of the record keeping associated with the incident report. Should an injury or death occur as a result of the response, documentation requirements need to be met. Just what documentation will be required varies, but the ISO must be aware of the requirements prior to the incident. A lack of knowledge concerning documentation requirements may cause loss of valuable information concerning the incident. This could result in loss of benefits to responders or their families or cause future litigation concerning the incident.

ACT AS A ROLE MODEL

If the ISO expects others to follow safe practices and procedures, he/she must set the example. The example should be established during training exercises and carry over to the incident scene. Safety is not something that starts with the response but must be an attitude that translates into forming safe habits. The ISO should be a role model and a mentor to others and should demonstrate positive leadership in creating such habits.

RECOGNIZING SAFETY CUES

The incident will present indicators that the ISO can use to recognize potential hazards. Smoke puffing through the mortar of a brick building or a sagging roof may be an indicator of impending building collapse. A gunshot incident at which no one knows where the gun or the shooter might be should alert responders to be on the lookout for a return visit from the shooter and the gun. Hastily established shoring at a trench rescue may be starting to bow. These are all examples of safety cues. Several other examples of safety cues can be seen in Figure 2-3. Safety cues are conditions or indications that the ISO needs to be aware of at an incident scene. These conditions or indications could be structural, unsafe acts by personnel, or unsafe conditions. The experienced ISO, when operating at an incident scene, will focus on these safety cues. Corrective action may need to be taken or modifications made to the action plan based on safety cues identified by the ISO.

In addition to the ISO's training and experience, he/she can develop skill in recognizing these safety cues based on what he/she has observed and learned from similar incidents. Familiarity with the department's or agency's day-to-day tactical operations will give the ISO the ability to recognize when the operations are different from those they typically perform and to determine if the differences are adversely affecting safety. Knowledge of any special operations that may be performed will allow the ISO to catch the safety cues that may be present. Knowing the level of training and experience of the responders will give the ISO a cue as to whether they are operating beyond their level of capability and possibly jeopardizing their safety.

FIGURE 2-3 The ISO must be able to identify safety cues and take the necessary action. Several safety cues are evident in this photo, including the colors of visible smoke, the location of smoke compared to the wind direction and visible fire, the lack of support for the porch roof, the amount of fire involvement, the location of power lines, the terrain around the building, and the snow on the ground. Photo by Robert Rosensteel, Sr., Vigilant Hose Company, Emmitsburg, MD.

Specialized Knowledge and Skills

To be as effective as possible, the ISO should have a thorough knowledge of the dangers and hazards the incident may present. Expecting an individual to have a detailed knowledge of the inherent dangers at every type of emergency incident is unreasonable. The ISO should be able to recognize personal limitations and not try to function in areas in which he/she is not knowledgeable.

For those incidents requiring specialized knowledge or skills, the ISO may need to seek assistance from someone who has the practical or technical expertise to deal with the emergency properly. This may be true particularly in some special operations that involve a number of agencies. A hazmat incident that results in multiple victims, leaking product, and a large-scale evacuation can present a multitude of problems for law enforcement, EMS, fire, public works, and a number of other agencies. Having technical assistance available for situations such as these can have a significant positive impact on the safety of both the responders and the members of the community involved.

At large-scale incidents or those covering a large area, it may not be practical for the ISO to handle all of the responsibilities demanded of the position. Due to incident complexity, the ISO may appoint others to assist in carrying out the duties. The ISO should assign specific areas of responsibility to ensure that all safety aspects of the incident are covered. The ISO must maintain an effective span of control and not get so overloaded that safety is jeopardized.

The ISO's Function in an ICS Organization

When the ISO is functioning with a formal ICS organization, he/she is part of the IC's command staff and reports directly to the IC (see Figure 2-4). He/She should get a briefing from the IC and determine what safety concerns are already addressed in the IC's action plan. The ISO must monitor those concerns and advise the IC of any present or potential hazards.

Although the IC has overall responsibility for safety at the incident, the ISO must aid the IC by focusing solely on the safety of the responders. The ISO should keep the IC updated on the conditions and should recommend any modifications that will address safety issues to the action plan.

Although the ISO may report directly to the IC, lines of communication must be established with others operating as part of the ICS organization. To get a thorough understanding of the tactical operations, the ISO likely will need to communicate and coordinate with the division, group, or sector supervisors. The operations chief, the medical unit leader, the rehabilitation manager, and other agencies might also be involved in the tactical operations.

As part of the ICS organization and the IC's command staff, the ISO needs to understand the IC's action plan and enforce the safety portion of the plan. The

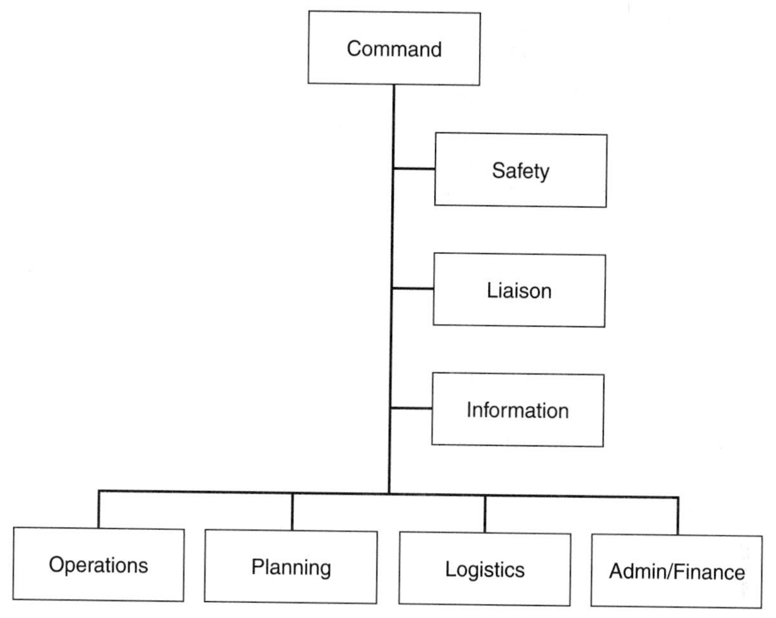

FIGURE 2-4 Incident Command System Chart.

ISO does this by first displaying a concern for responder safety. The actions the ISO takes to display that concern may include enforcing SOPs and recognized safety practices

Along with accepting the role of ISO, the individual who fills the position must also accept the responsibility the IC has delegated for providing for the safety of the responders. It is not a responsibility to be taken lightly. It requires knowledge, personal and professional skills, and a strong commitment to provide for the safety of the responders.

The Health and Safety Officer (HSO)

According to NFPA 1500, *Standard on Fire Department Occupational Safety and Health Program,* every fire department, regardless of size, must have an individual assigned to the duties of health and safety officer (HSO). This individual was once known as the fire department safety officer (FDSO), but the title has been changed to more accurately fit the role of the position and the diversity of the agencies an HSO may work for. An HSO also is strongly recommended for other emergency response organizations, such as EMS departments. The HSO must be an officer within the agency, is most often appointed by the fire chief or head of the agency, and reports to the chief or the chief's designee. Status as an officer will help the HSO provide the needed safety emphasis for the agency. The HSO's most significant responsibility is managing the agency's safety program. Ultimate responsibility for the safety of responders still rests with the fire chief or the head of the organization, but the HSO can help the fire chief ensure that responders are safe by placing an emphasis on safety. There is only one HSO per response agency.

The HSO is a departmental level, primarily administrative officer responsible for coordination of safety and wellness aspects as they apply to organizational activities. This includes emergency and nonemergency operational and day-to-day activities. The HSO guides departmental policy as it applies to occupational safety practices and member welfare issues. The HSO is responsible for interpretation of rules, regulations, and standards and their application as they relate to organizational activities.

NFPA 1521 outlines requirements for the training, certification, and functions of the HSO. The fire department HSO must be trained to NFPA Fire Officer Level I according to the requirements of NFPA 1521, *Standard for Fire Department Safety Officer.* All safety officers, including the HSO, must have a knowledge of fire fighting and its hazards, rules, and regulations that affect fire department operations, such as NFPA standards, departmental standard operating procedures (SOPs), safety management, and physical fitness. The majority of the HSO's activities will be performed in a nonemergency setting; however, the work of the HSO has a major impact on emergency operations. The HSO may act as an incident safety officer (ISO) and provide safety supervision on emergency scenes.

Basic Duties and Responsibilities of an HSO

Some of the major duties of the HSO include the following.

- Development of SOPs for high-risk activities such as highway operations (see Figure 2-5), emergency driving, wildland and structural fire fighting, and the use of protective clothing and equipment such as self-contained breathing apparatus (SCBA)
- Development and delivery of safety training courses for members of the fire departments
- Development and management of injury and exposure documentation systems
- Input into all fire department pre-emergency, emergency, and nonemergency operations to ensure that the safety of responders is considered at all times
- Development and support of appropriate physical fitness programs for responders
- Analysis of fire department and national firefighter injury and death studies to develop means of preventing the recurrence of such incidents
- Help with the research, purchase, and use of protective clothing and equipment
- Knowledge of national standards for safety equipment and operational safety

FIGURE 2-5 The HSO's efforts will have a major impact on emergency response and operations through the development of standard operating procedures, training in safe operations, and the purchase/use of protective equipment and safety devices. Photo by Steve Weissman, Fairfax County (VA) Fire and Rescue Department.

The Relationship of the Health and Safety Officer and the Incident Safety Officer

The duties of the HSO and the ISO are similar in many ways: They both focus on firefighter safety. The primary difference between the HSO and the ISO is the setting for their work. The work of the HSO is generally pre-emergency or nonemergency in nature; the work of the ISO most often is performed at the scene of an emergency.

If the HSO performs safety duties at the scene of an emergency, he/she is referred to as the ISO. The HSO is a single individual who is usually appointed by the fire chief to focus on responder safety. Because there is only one HSO, he/she cannot possibly be available to respond to every emergency incident and perform as the ISO; however, the safety of responders must be addressed at every incident. If the HSO is unavailable to act as the ISO, the IC will appoint another member of the agency to perform the functions of the ISO.

In some very large agencies, on-duty ISOs may be assigned. For most agencies, the ISO may be any member of the agency who is assigned at the scene of the emergency.

The ISO performs a set of standard tasks on the emergency scene. If an assigned ISO is not the HSO, the same tasks still must be performed. The HSO should ensure that other members of the agency are trained to a level that allows them to perform as an ISO.

In situations where the IC appoints someone other than the HSO to perform ISO duties, the HSO and the responders working on the scene of the emergency depend on the ISO to monitor safety. At conclusion of the incident, the ISO should tell the HSO of any unusual problems or situations encountered at the emergency scene. The HSO can use this information to develop additional safety procedures. For example, existing safety procedures may need to be revised, or other problems may need to be addressed through the use of protective clothing and equipment.

At complex or large emergencies, more that one ISO may be needed to monitor the safety of responders. The initial ISO should compile information from all of the other ISOs at the incident and provide any information on problems or unusual situations to the HSO.

SUMMARY

The basic duty and responsibility of the ISO is to protect the safety of the responders. The ISO as an effective risk manager can protect the safety of responders. The ISO needs a knowledge of safety concerns at incidents, must know his/her limitations and when to get help, and be able to work as an integral part of an ICS organization.

The HSO provides a safety focus and is constantly on the lookout for safety issues within the organization. The HSO also looks at agency operations from a point of view different from that of other officers in the agency (safety versus operations). The ISO can take a safety-oriented view of emergency operations. Responders tend to focus closely on the tactical objectives of the incident.

3

Laws, Regulations, and Standards

The health and safety officer (HSO) must have a working knowledge of the laws, standards, and regulations that affect safety and health. This position serves as a clearinghouse for information relating to the development and revision of laws, standards, and regulations that affect the department, especially safety and health issues.

The incident safety officer (ISO) must also deal with a variety of regulations, standards, and policies on a regular basis. It is important to realize that these regulations originate in different ways and affect different organizations in different ways. This chapter explores some of the regulations that currently apply to the HSO and the ISO.

As the United States became more industrialized and technology improved at the turn of the twentieth century, the workplace was affected by the changes. Instead of a business, company, or industry focusing on productivity at any cost, safety mandates and workers' rights became more important. Previously, most employers had focused strictly on productivity and profit with little or no regard to workers' safety and health. Safeguards for equipment and machinery did not exist. If an employee was injured, disabled, or killed, the employee was replaced with little or no compensation to the employee or relatives. Most employers were exonerated of any criminal or civil penalties because no laws relating to compensation or employer liability existed at the time.

As technology increased, so did emphasis on employees' rights and their safety and health. This was very evident with the development of child labor laws. These laws protected children's rights relating to employment requirements and hours worked.

Employee safety and health rights also began to improve. Safety and health issues such as workplace fire prevention and protection, safety guards for equipment and machinery, and medical treatment for occupational injuries were addressed by the employer. Workers' compensation for occupational injuries and accidents developed. This process culminated with the passage of the Occupational Safety and Health Act of 1970.

Several national standards, especially NFPA 1500, *Standard on Fire Department Occupational Safety and Health Program,* have had a major influence on member safety and health within the past several years. NFPA 1500 is the *only* standard that comprehensively addresses necessary components for a fire service occupational safety and health program. The importance of complying with this standard is indicated by the reduction in firefighter fatalities, the reduction in the number and severity of accidents and injuries to members, and improved health maintenance and well-being of firefighters. This is a proactive approach to reducing risk for personnel and reducing liability to the department.

Laws

The safety and health process has continued to improve over the past 20 to 25 years, especially for emergency services personnel. Protective clothing, protective equipment, and operational procedures have improved greatly through the enactment of particular laws.

Statutory Law

Statutory laws apply to civil and criminal matters and are enacted by a body legally authorized to legislate such requirements. These governing bodies can be state legislative assemblies, which affect only one particular state, or the United States Congress, which affects all states and territories. After the bill has fulfilled the necessary requirements and has been passed by the legislative body and signed by the executive, it becomes law.

One example of a statutory law is the Superfund Amendments and Reauthorization Act (SARA) of 1986. Section 126 requires the Environmental Protection Agency (EPA) to issue an identical set of regulations covering anyone not covered by 29 CFR 1910.120, *Hazardous Waste Operations and Emergency Response* (HAZWOPER). States that operate under federal OSHA regulations (rather than under state OSHA plans) are required to comply with the requirements of the EPA, which are identical to OSHA's requirements. The importance to emergency response personnel is that both require the use of an incident management system; use of an incident safety officer (ISO) at hazardous materials incidents; and use of a health monitoring process for employees exposed or potentially exposed at a hazardous materials incident.

The Ryan White Comprehensive AIDS Resources Emergency Act (CARE) of 1990 has provisions that require emergency responders to be notified if they have been exposed to a communicable disease during treatment of a patient. Although body fluid exposures may be obvious to a responder, exposure to an airborne disease may not be as evident. The notification process requires that the testing source or agency, such as a hospital, notify the affected employee directly. The Ryan White CARE Act protects the confidentiality of the affected employee.

Each state or commonwealth has legislative assemblies that pass laws affecting emergency service personnel. The process is similar to the action Congress takes in the development of a law. A bill is introduced to this lawmaking body; if the legislation is approved and signed by the governor, it becomes law.

Examples of laws affecting fire and EMS departments are those that govern the emergency response of fire and EMS vehicles. Most jurisdictions have requirements with defined responsibilities for operating vehicles under emergency conditions. This would include emergency vehicle operations when passing through intersections with traffic lights, stop signs, or yield signs or when encountering a school bus that is stopped and discharging passengers.

Development of an annual vehicle inspection program for all fire apparatus has become law in several states over the past few years. Due to the number of firefighter fatalities from vehicle accidents caused by poor vehicle maintenance, state legislatures mandate that all fire apparatus be inspected annually by a certified inspector. The Commonwealth of Virginia passed such a law, which became effective in July 1992, requiring all fire apparatus to be inspected annually.

Case Law

Case law is different from statutory law in one very large respect: Case law is not voted on by a legislative body but comes either from legal precedent or judicial decision in particular cases. In emergency services, these judicial decisions usually follow a civil suit. These decisions affect responders by changing requirements or procedures rapidly, usually with little or no implementation time. A statutory law may not take effect for years.

Depending on the court that hears the case, the decision may be either nationwide or statewide in scope. The case law of the country is always subject to change. Similar cases heard in different state courts can have opposite findings. Sometimes, the decisions of the courts are later reversed by the same or a different court.

In *Whirlpool Corp. v. Marshall* (100 S. Ct. 883, 1980), it was determined that an employee has the right to choose not to perform an assigned task because of a reasonable apprehension that death or serious injury may result. Specifically, this right is applicable when the worker believes there is no less drastic alternative than refusing to work. The courts in this case found for the worker and upheld this right. This is one example where a decision by a court may have had an effect on emergency responders, although the case in question did not involve emergency responders.

It is important to perform legal research of both federal and state cases to determine which cases affect you. It would be appropriate to obtain assistance for this research from people who have more legal experience than you.

Regulations Issued by Federal Administrative Agencies

The U.S. Congress has created different federal administrative agencies. These agencies issue rules, regulations, or orders that have as much authority as statutory laws. You have been affected by these rules, but you probably did not give it much thought. Have you ever noticed that all school buses are the same color?

The Department of Transportation (DOT) is responsible for that. How about rules concerning airline safety? Airline rules and regulations come from the Federal Aviation Administration (FAA).

One agency of particular interest to emergency responders is the Occupational Safety and Health Administration (OSHA), because its rules govern worker safety. OSHA, a branch of the Department of Labor, was created in 1970 to provide written requirements for ensuring occupational safety and health standards for employees.

Under the Occupational Safety and Health Act of 1970, federal OSHA has no direct power to ensure that state and local governments comply with safety and health standards for public employees. OSHA law does permit other methods to be used in order to provide maximum protection of public employees' safety and health.

Twenty-five states and two territories have established and maintain an effective and comprehensive occupational safety and health program for public employees. This state plan must meet or exceed the requirements of federal OSHA. OSHA gives a state plan 6 months from the publication date of a final standard to adopt a similar standard. All state, county, or municipal fire departments in any of the states or territories having an OSHA plan agreement in effect have the protection of the minimal acceptable safety and health standards mandated by federal OSHA.

Several OSHA standards affect department personnel. 29 CFR 1910.120, *Hazardous Waste Operations and Emergency Response* (HAZWOPER), and CFR 1910.1030, *Occupational Exposure to Bloodborne Pathogens,* mandate requirements for maintaining an employee's medical records for 30 years past retirement date (29 CFR 1910.20, *Access to Employee Exposure and Medical Records).*

Other OSHA standards that the HSO should be familiar with are 29 CFR 1910.95, *Occupational Noise Exposure,* 29 CFR 1919.133, *Eye and Face Protection,* 29 CFR 1910. 134, *Respiratory Protection,* and 29 CFR 1910. 156, *Fire Brigades.*

Consensus Standards

A consensus standard is established by an authority or by general consent as a model or example to be followed. These standards are usually developed after an incident, but they can be developed proactively to address a safety concern. Nonregulatory groups and associations develop the standards with suggestions from members who have asked to participate in the standards-making process. The group that publishes the most fire service consensus standards is the National Fire Protection Association (NFPA). NFPA has several standards applicable to EMS, as well.

The American Ambulance Association (AAA) currently is using consensus standards to grant certifications to different ambulance companies. Although these certifications are not mandatory, the association is using them as a method of self-regulation. ATSM also develops and publishes EMS-related standards.

Consensus standards usually are not mandatory, but the legal status of consensus standards varies from state to state. The standards are used extensively in litigation and, in the event of a lawsuit, you may be asked to defend a failure to follow the standard.

Standards-making organizations or associations have guidelines that dictate the procedures for developing standards. The structure of the committees must be balanced so that one interest or group cannot dominate the committee. The number of members is another consideration. The process of incorporating comments or suggestions into committee standards; how a new document is developed; the length of the process; the revision process; and how long it takes to revise a standard are just a few of the guidelines that govern the standards-making process (see Figure 3-1).

The poor safety record of the fire service over the years has been the leading impetus for developing standards. These standards have addressed personal

FIGURE 3-1 Many fire and EMS department safety officers have participated in the standards-making process, either by becoming members of technical committees or by reviewing and commenting on proposed standards related to fire service occupational safety and health. Photo by Gerry Suftko, Mesa (AZ) Fire Department.

protective equipment and clothing, apparatus, hazardous materials, infection control, and other pertinent issues.

Consensus standards are not mandatory unless officially adopted by public authorities with lawmaking or rule-making abilities. Once a legislative body adopts and makes a consensus standard a law in whole or in part, the consensus standard becomes a mandatory requirement in that jurisdiction.

National Fire Protection Association (NFPA) Standards

Since 1896, the National Fire Protection Association (NFPA) has been the leading nonprofit organization in the world dedicated to protecting lives and property from the hazards of fire. NFPA is noted for its involvement in fire prevention and education programs and the standards-making process. NFPA has developed several of the most well-known and widely used standards, such as the *National Electrical Code* and the *Code for Safety to Life from Fire in Buildings and Structures.* The NFPA publishes over 270 nationally recognized codes and standards.

In 1986 at the NFPA annual meeting in Cincinnati, Ohio, the NFPA membership passed NFPA 1500, *Fire Department Occupational Safety and Health Program.* This standard has affected the safety and health of the fire service as has no other. The NFPA Technical Committee on Fire Service Occupational Safety and Health and the NFPA Technical Committee on Fire Service Occupational Health and Medical are responsible for the development of several standards that address specific safety and health issues. The original intent was to expand each chapter of NFPA 1500 into its own document. These newer standards include the following.

- NFPA 1521, *Standard on Fire Department Safety Officer*
- NFPA 1561, *Standard on Fire Department Incident Management System*
- NFPA 1581, *Standard on Fire Department Infection Control Program*
- NFPA 1582, *Standard on Medical Requirements for Fire Fighters and Information for Fire Department Physicians*

NFPA 1500

The NFPA intent of the Technical Committee on Fire Service Occupational Safety and Health was to develop a user standard addressing various safety and health interests. NFPA 1500 serves as a voluntary consensus standard detailing the needed requirements of a complete fire department safety program. Once the authority having jurisdiction (AHJ) adopts the standard, it then becomes a mandatory standard. The *authority having jurisdiction* is defined as "the organization, office, or individual responsible for approving equipment, an installation, or a procedure." NFPA 1500 meets or exceeds the criteria listed in subpart L of the OSHA requirements.

This document contains the minimum requirements for a fire department occupational safety and health program. NFPA 1500 is intended to reduce the number and severity of accidents, injuries, and exposures. Another important issue is ensuring that all persons abide by this standard, whether the person is a career, part-paid, or volunteer firefighter. The technical committee felt safety should incorporate any person regardless of his or her affiliation. The term *member* is used throughout the standard to define a person involved with a fire department organization.

NFPA 1500 states that a fire department organization should develop and maintain a written safety and health program. With this program, the organization must develop goals and objectives to prevent and eliminate accidents, occupational illnesses, and fatalities. A fire department HSO must be appointed and be responsible for the occupational safety and health program. The duties and responsibilities of the HSO are defined in NFPA 1521, *Standard on Fire Department Safety Officer.* The occupational safety committee will conduct research, develop recommendations, and study and review matters pertaining to occupational safety and health within the fire department.

The NFPA also publishes standards that address protective clothing and equipment; these are the manufacturers' design criteria for protective clothing and equipment. As applicable clothing and equipment are ordered, the department should stipulate that they meet these specifications.

Clothing and equipment that must meet NFPA standards include the following.

- Structural firefighting clothing and equipment, such as turnout coats, turnout pants, helmets, boots, and gloves
- Protective equipment, such as self-contained breathing apparatus (SCBA), personal alert safety systems (PASS), and rope
- Protective clothing for personnel involved in wildland, hazardous materials, and crash, fire, and rescue operations
- Protective clothing for personnel engaged in emergency medical operations

NFPA has design criteria for various types of fire apparatus and equipment. These criteria cover the following:

- Manufacturers' specifications for pumpers, aerial ladders and platforms, and tankers
- Design specifications for hose, ground ladders, and nozzles
- The annual testing criteria for apparatus and equipment to ensure safety in operations

NFPA technical committees develop professional qualifications and competency standards for firefighters, fire officers, driver/operators, and fire instructors.

In addition, standards relating to live fire training, training centers, and training reporting procedures have been developed. NFPA 1403, *Standard on Live Fire Training Evolutions in Structures,* was developed because of numerous firefighting fatalities during live fire training exercises.

ASTM (formerly American Society for Testing and Materials)

ASTM is a private, nonprofit organization that develops standards for materials, systems, products, and services. It was founded in 1898 to provide services for a variety of disciplines.

Standards for emergency medical services were developed by ASTM's F-30 Committee on Emergency Medical Services. These standards include *Standard Practice for Training the Emergency Medical Technician (Basic)* (F-1031) and *Standard Guide on Structures and Responsibilities of Emergency Medical Services Systems Organizations* (F-1086).

ASTM D 3578, *Standard Specification for Rubber Examination Gloves,* 1991, includes requirements for sampling to ensure quality control; water-tightness testing for detecting holes in gloves; physical dimension testing to ensure proper fit of the gloves; and physical testing to ensure that the gloves do not tear easily.

Testing methods for personal protective clothing include ASTM F 739, *Test Method for Resistance of Protective Clothing Materials to Permeation by Liquids or Gases Under Conditions of Continuous Contact,* 1991, and ASTM F 1052, *Standard Practice for Pressure Testing of Gas-Tight Totally Encapsulating Chemical-Protective Suites,* 1987. ASTM F 1052 is referenced in NFPA 1991, *Vapor-Protective Suits for Hazardous Chemical Emergencies,* as a method of testing the gas-tight integrity of the respective protective suit.

Standard Operating Procedures

Standard operating procedures (SOPs), often referred to as standard operating guidelines or SOGs, are written policies developed by a department that specify methods for activities performed by members of the department. These procedures affect only the operation of the department that writes and adopts them. The requirements of these procedures must be based on recognized laws and regulations. The department must meet or exceed the requirements. The SOPs should also conform to any applicable standards that may address the same issues.

Reasons for These Laws, Regulations, and Standards

All laws, regulations, case law, and standards are designed with the safety of the responder in mind. At times, it may seem that the procedures hinder the responder at an emergency, but that is when the need for safety is most critical. As responders speed up their actions, the chances of injury increase.

By following all rules and regulations, the liability to both the responder and the agency is reduced. As an ISO, you have the task of ensuring that all responders act in a safe manner. Because you may know the responders personally, it is important to stress that the rules are there to protect them.

Uninjured responders remain available for other calls. This helps lessen the problems inherent in running without a full complement of personnel. It is noticeably cheaper to run an organization that prevents injury. Paid departments avoid overtime payments, because responders are present for duty. In volunteer organizations, insurance premiums may be reduced. In both cases, the psychological toll of treating injured fellow responders is lessened. The toll on responders who have to treat other responders with serious injuries is high. Additionally, the rate of injuries to responders after a fellow responder has been killed or seriously injured tends to increase.

What Do the Regulations Cover?

The different regulations cover a variety of issues, including work practices, scene safety, and responder welfare. All of these regulations are designed to benefit the emergency responder.

INFECTION CONTROL

Infection Control Officer

Under the Ryan White Act, each agency is required to appoint a designated officer to handle infectious disease issues. This officer is usually referred to as the infection control officer. The agency also is required under OSHA 29 CFR 1910.1030, *Occupational Exposure to Bloodborne Pathogens,* to have a written exposure control plan. In most agencies, the infection control officer also administers the infection control plan. This is also covered in NFPA 1581, *Standard for Fire Department Infection Control Program.*

An ISO may not deal with the infection control plan administratively; however, the ISO will have to see that the plan is followed in the field (see Figure 3-2). In order to monitor compliance with the plan effectively, the ISO must work with the infection control officer. The ISO must maintain an ongoing working relationship with whoever is designated and be familiar with their responsibilities.

As ISO, you should review your agency's plan and identify those areas of responsibility that overlap with those of the infection control officer. By meeting regularly with the infection control officer, you both can develop an effective strategy to maintain a high compliance ratio in the field.

Post-Exposure Procedures

You must be familiar with the post-exposure part of the infection control plan. Because you will be in the field, you may either be called to the scene or notified of a responder's exposure. The infection control officer may be an administrative and not a response position; you may be responsible for carrying out the post-exposure part of the plan.

Your primary concern should be the prompt care of the exposed responder, but you must remain aware of the need for patient confidentiality. As with any incident, there may be laws governing the confidentiality of both the responder's and the original patient's medical records. Maintaining the confidentiality of certain records is necessary for the protection of both responders and patients.

Follow-up and any concerns you may have concerning patient confidentiality can be discussed with the infection control officer, who should be versed in this area. If that officer cannot answer your questions, then you should be directed to someone who can.

The first step in the post-exposure process is notification. You may be notified by seeing the exposure, getting a telephone call from the responder or the hospital, or from talking with the responder's supervisor.

After you have been notified, the next step in the process is verification. You should refer to your department's infection control plan to verify that the exposure can be defined as a "true exposure" and to determine whether the responder needs immediate medical attention.

If the responder needs medical care, ensure that it is obtained according to the department's SOP. Included in the care will be follow-up and counseling, which the infection control officer should arrange.

FIGURE 3-2 Fire and EMS personnel are at high risk of exposure to communicable diseases through bloodborne and airborne transmission. Laws, regulations, and standards require the use of appropriate personal protective equipment in order to minimize this risk. Should an exposure occur, protocols for notification, documentation, and testing/treatment/follow-up are also required. Photo by Orlando Dominguez, Brevard County (FL) Fire/Rescue.

Your last step should be to ensure that your department's required paperwork is completed and filed appropriately. A call should be made the next day to the infection control officer so you can give a short briefing on the incident.

Personal Protective Equipment (PPE)

You should be experienced in using PPE for infection control. In the field, you will be expected to lead by example and use the equipment when required. If a responder at an incident is having difficulty with a piece of equipment, you should assist. Sometimes, all that is required is for you to detail a responder to remain at an access point to hand out equipment, especially gloves.

After an incident, a responder (in proper PPE) must clean up the scene. It is unprofessional to leave disposable equipment like gloves and bandages, which may have been contaminated with body fluids, at the scene. In some states, a department leaving bloody gloves behind can be fined for improper medical waste disposal. You should ensure that all potentially infectious material is disposed of properly and that personnel are trained to wear proper PPE when gathering medical waste.

Proper disposal includes the station. The infection control officer should be managing this aspect of the infectious waste stream; however, an HSO can always make a spot check and let the infection control officer know when there is a problem. When you make spot checks, any area that is used for cleaning of infectious nondisposable equipment also should be checked, as well as proper use of PPE by personnel who clean the equipment.

HAZARDOUS MATERIALS

Rules and Regulations

As mentioned earlier, the area of hazardous materials response is regulated by OSHA 29 CFR 1910.120, *Hazardous Waste Operations and Emergency Response.* This rule is binding on all responders, and it is not permissible to say "We do not handle hazmat calls." OSHA's definition of *emergency response* is "a response effort by employees from outside the immediate release area or by other designated responders (i.e., mutual-aid group, local fire departments, etc.) to an occurrence that results, or is likely to result, in an uncontrolled release of a hazardous substance." You can see that this definition is intended to cover any responder who handles emergency calls.

Some states do not follow OSHA rules; those states have adopted laws that at least equal the OSHA rules but generally cover only private employee, not emergency, responders. The EPA, under section 126 of the Superfund Amendments and Reauthorization Act (SARA) of 1986, is required to issue a set of regulations that are identical to OSHA's rules. The EPA regulations cover anyone who is not covered by OSHA (see Figure 3-3).

Both OSHA and the EPA require the use of an incident command system (ICS). It is imperative from a safety standpoint that an ICS be employed at a hazardous materials incident so that responders can be tracked and monitored. In the

FIGURE 3-3 Hazardous materials emergency response is covered specifically by OSHA and EPA regulations, which together apply to all fire and EMS departments. Some of the requirements include the use of ICS, including an incident commander and safety officer, and appropriate training and equipment for those involved in emergency response to hazardous materials releases. Photo by G. Emas, Edmonton (Alberta) Emergency Response Department.

event of a rapid negative outcome at an incident, the ISO is responsible for ensuring that all responders are accounted for. The rules also require the on-site presence of the ISO.

The function of the ISO should be delineated clearly in your agency's SOP. Law also mandates the use of SOPs at hazardous materials incidents. It would be prudent for you to review your role in the SOP and recommend any safety-related changes prior to an incident. Included in the law is the need for a site-specific safety plan. As the ISO, it becomes your responsibility to create a plan. This plan should be in writing and on a form approved by your agency. A sample form can be found in appendix F.

By ensuring that all applicable aspects of the rules are met prior to an incident, you can reduce the risk faced by your organization. When the risk is reduced, your liability also will be reduced. Responders must understand their roles and the limitations of their training.

Technical Experts

Safety officer training alone will not provide all of the knowledge needed to be the only ISO on the scene of a hazardous materials release. You should know your limitations and be able to gain assistance from a hazardous materials technical expert. If your local hazardous materials team responds with a hazardous material safety officer, that officer would be the appropriate person to give you advice and work with you. If the team does not have such a person, then you will have to review the safety options with the hazardous materials team leader.

CONFINED SPACES

Confined-space emergencies can present responders with a unique and dangerous type of response. Sometimes, the responders do not realize the danger they are subjecting themselves to when they perform confined-space rescues. In recognition of this, OSHA has established a rule for safety during confined-space operations. This rule is 29 CFR 1910.146, *Permit-Required Confined Spaces.*

Among the law's provisions are the definition of a confined space; what atmospheres are hazardous; and the use of a personnel tracking system. The law also mandates the use of a clean air supply during rescues. Confined-space entry permits must be obtained prior to personnel entering the space; an example of such a permit can be found in appendix G.

Just like the hazardous materials laws, this law is intended to protect the responder. It places the safety of the responder first and allows the agency to limit its risk by mandating that procedures to handle confined-space emergencies be in place prior to an incident.

Confined-space rescue is another area where the use of technical experts is crucial. If your agency seldom responds to confined-space emergencies, you should locate someone in your area who is available to assist and develop a working relationship prior to an incident.

DEVELOPMENT OF WRITTEN SAFETY AND HEALTH PROCEDURES

As with any procedure, if a department expects all personnel to understand the requirements of the department's occupational safety and health program, the program must be a written one. This written program will provide the necessary components for the department's safety and health program. As part of the program, the duties and responsibilities of the HSO and occupational safety and health (OSH) committee should be defined.

NFPA 1500 Section 2–3.1 states, "The fire department shall adopt an official written departmental occupational safety and health policy that identifies specific goals and objectives for the prevention and elimination of accidents and occupational injuries, exposures to communicable diseases, illnesses, and fatalities. It shall be the policy of the fire department to seek and to provide an occupational safety and health program for its members that complies with this standard for its members."

Health and Safety Officer (HSO)

Every department should recognize the need for an HSO; the organizational structure of a department will dictate job functions of the HSO. In a medium or large department, an HSO position may be established and operate on a daily basis and/or a shift schedule. In a small department, the OSH committee may assume the duties in lieu of an HSO, or the position may be assigned as a part-time position. Regardless of the size of the department, there should be an OSH committee.

The HSO will be responsible for managing the daily operations of the safety and health program. Duties will be defined by department policy, position description, or within the safety and health program. NFPA 1521 lists prescribed duties for the HSO.

Occupational Safety and Health (OSH) Committee

If the OSH committee has the primary function of managing the safety and health program, the duties and responsibilities may be divided among committee members. One member may be assigned to apparatus safety, another to training safety, and another to facility safety until all the functions have been assigned. NFPA 1500 provides guidelines for the OSH committee.

PERIODIC REVIEW AND REVISION PROCESS

A department must have procedures in place to evaluate, review, and revise procedures routinely. The HSO and/or the OSH committee will be responsible for this process.

NFPA 1500 Section 2–3.2 states, "The fire department shall evaluate the effectiveness of the occupational safety and health program at least once every three years. An audit report of the findings shall be submitted to the fire chief and to the members of the occupational safety and health committee." A qualified individual from outside the department should conduct this audit (see Figure 3-4); outside evaluators provide a different perspective, which can be constructive.

The job functions and duties of the HSO or the OSH committee must be reviewed and revised routinely, based on new assignments or responsibilities. This determines whether the HSO is concentrating on the primary functions of the position. If an HSO position is established, the OSH committee still must function as part of the department's safety and health process.

SUMMARY

The primary responsibility of an employer is to provide a safe and healthy work environment. The tangible benefits are the reduction of risk to the employee and the decrease in liability for the department. Unfortunately, the fire service tends to confront these issues from a negative perspective. The positive considerations should be addressed, as well as how to improve the overall operations of the department by reducing risk to personnel and limiting or decreasing liability.

FIGURE 3-4 A qualified individual from outside the fire or EMS department should be used to evaluate a department's occupational safety and health program and ensure that all applicable laws, regulations, and standards are addressed in the program. Photo by Ed Roberts, Dothan (AL) Fire Department.

The fire chief has the ultimate responsibility for safety and health in the department. So that the process can be properly managed, the fire chief appoints or designates an HSO to be the manager of the safety and health program.

Ensuring compliance with all applicable laws, standards, and regulations develops and promotes a positive image inside and outside the department. Complying with safety and health standards demonstrates a responsibility for safety and health on the part of all personnel. It sends a message that the organization cares about employees and their well-being. An organization that has effective operating procedures provides a more efficient and productive operation on a daily basis. Personnel are provided with the essential requirements to function in their assigned positions.

A department develops policies based on laws, standards, and regulations to provide the necessary information to establish procedures. For example, NFPA 1500 provides the necessary components to establish a comprehensive safety and health program. Prior to the development of NFPA 1500, no other law, standard, or regulation addressed safety and health needs for the fire service in such an inclusive manner.

The HSO must develop a network that provides information on laws, standards, and regulations. As these are revised or developed, the HSO can include this information in the department's procedures.

By understanding the different regulations, standards, and policies, the ISO can effectively perform his/her job and provide a safer working environment.

No one ever called an emergency response agency because everything went right. You are not operating in vacuum. Develop a support system for yourself. This support system will assist you in the performance of your job functions. An ISO must work as part of the overall incident management team, advising the IC on safe work practices and legal mandates.

4

Record Keeping and Documentation

The health and safety officer (HSO) is primarily responsible for collecting, reviewing, and analyzing accident and injury reports. The HSO serves as the primary player in an ongoing or lengthy investigation. The HSO and the incident safety officer (ISO) must work together closely to ensure that reports are properly initiated and completed.

The ISO manages initial investigations. If the investigation cannot be completed by the ISO due to time or length of the investigation, the HSO must assist. Generally, the ISO is the person on site who immediately starts an investigation to document an incident properly. The ISO also can assist other department personnel in completing their reports.

Firefighter Fatality and Injury Data

For the last decade, an average of approximately 100 firefighters have been killed in the line of duty each year, and about 100,000 were injured. These numbers are viewed by some as staggering, but they are much improved over past decades. These statistics are shown graphically in Figures 4-1 and 4-2.

Statistics show trends that indicate improvement, or regression, in firefighter fatalities, accidents, and injuries. These statistics can be used to influence operations and affect the safety of department personnel.

In order to understand the safety process, the fire service must focus on where personnel accidents and injuries are occurring. This gives an emergency services organization the parameters to develop a strategy to reduce the number and severity of these incidents.

National Statistics

In 1974, the NFPA began to collect data on firefighter fatalities in the United States. The United States Fire Administration (USFA) also conducts extensive analyses of patterns and trends in specific areas of the firefighter fatality problem. The National Institute for Occupational Safety and Health (NIOSH) is congressionally mandated to investigate firefighter fatalities.

NFPA publishes an annual injury survey that estimates the number of firefighter injuries. These are not complete data, like the fatality study, because fire

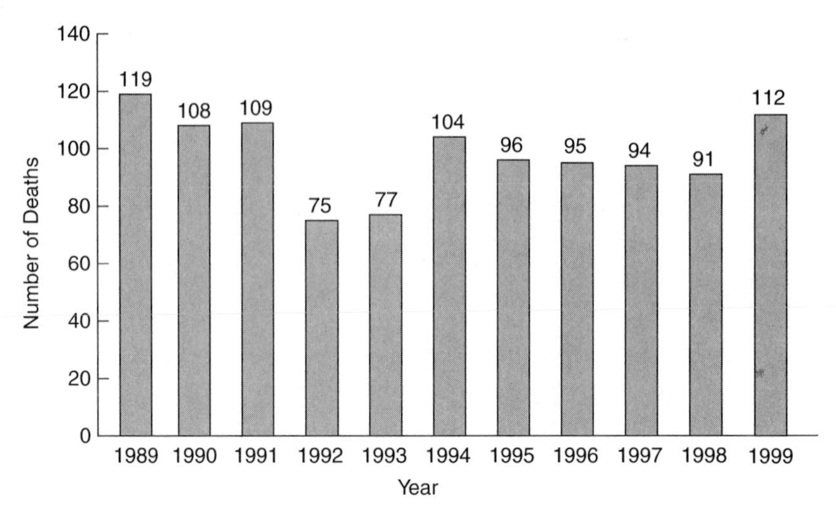

FIGURE 4-1 On-duty firefighter fatalities, 1989–1999. *Source:* Firefighter Fatalities in the United States, 1999, USFA, p. 7.

services' input is voluntary. The statistical data for fatalities and injuries consistently show that more firefighters' deaths and injuries occur at emergency scenes than at any other job function performed.

The International Association of Fire Fighters (IAFF) produces an annual firefighter fatality and injury study. This report includes statistics from U.S. and Canadian departments that are members of the IAFF. Report data generated by the IAFF include statistics on lost work time, disability retirements resulting from accidents and injuries, EMS injuries, and physical fitness injuries.

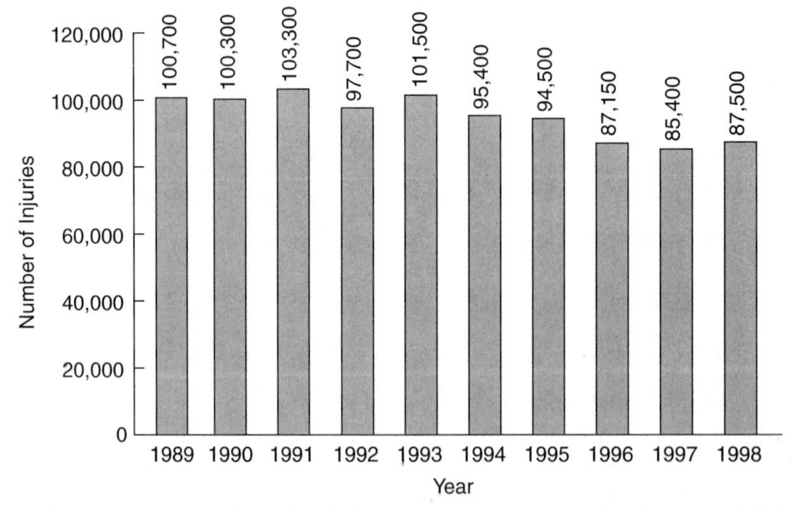

FIGURE 4-2 On-duty firefighter injuries, 1989–1998. *Source: NFPA Journal*, November/December 1999, p. 48.

Department Statistics

Each emergency services organization should track incident statistics relating to death, injury, or illness to department personnel. This information can provide valuable data for improving personnel safety and reducing risks.

Record keeping and documentation are especially important for health problems associated with exposures to hazardous materials or communicable diseases. Due to the extended lead time associated with chemicals and communicable diseases, the health effects or illness may not appear for years. If there is incomplete or no documentation of a health exposure, a member may be refused job-related medical coverage.

Agencies responsible for medical and health coverage for personnel have very strict guidelines for reporting and documenting accidents, injuries, and illnesses. Failure to comply with these guidelines can be costly to the department and to the individual member.

Importance of Record Keeping

Occupational Safety and Health Administration (OSHA)

OSHA regulations affect public fire departments in 23 states and territories that have state OSHA programs. The remaining states come under federal OSHA guidelines, which do not include municipal or volunteer fire departments, although federal, military, and private fire departments are required to meet OSHA standards.

To ensure that employers record and maintain injury data relating to employees, OSHA has mandated requirements for record keeping. Each year, an employer must publish and post data on a specific OSHA form called the "OSHA Log 200."

These regulations require the employer to document the following.

- Occupational death
- Nonfatal occupational illness
- Nonfatal occupational injury

For occupational injuries, one or more of the following must have occurred.

- Loss of consciousness
- Restriction of work or motion
- Transfer to another job
- Medical treatment other than first aid

Other information required for OSHA Log 200 includes the employee's name, job title, and work assignment; type of injury or illness; type and extent of medical treatment; and length of lost work time, if any. OSHA Log 200 must be posted from February 1 to March 1 of each year.

Although only a consensus standard, NFPA 1500, *Standard on Fire Department Occupational Safety and Health Program,* states in paragraph 2-7.1, "The fire department shall establish a data collection system and maintain permanent records of all accidents, injuries, exposures to infectious agents and communicable diseases, illnesses, or deaths that are or might be job related." NFPA 1521, *Standard for Fire Department Safety Officer,* states in paragraph 3-6.1: "The fire department shall maintain records of all accidents, occupational deaths, injuries, illnesses, and exposures in accordance with Chapter 2 of NFPA 1500, *Standard on Fire Department Occupational Safety and Health Program.* The health and safety officer shall manage the collection and analysis of this information."

Most often, the HSO is responsible for maintaining OSHA Log 200. Municipal or county governments may assign this duty to a safety administrator or risk manager. These individuals would maintain the records for all city or county employees.

Occupational Insurance

In order to better protect and serve its employees, and to protect its investment from a time and monetary standpoint, an organization must provide a means of medical coverage. This can be done in a variety of ways, depending on the type of organization.

A municipal department may be covered by a state or local workers' compensation program. Many municipal organizations are self-insured, which means that they administer their own insurance programs. Due to the exorbitant costs of liability insurance, organizations have found it less expensive to operate their own programs. For the most part, this is done through the finance department or division of the organization, commonly referred to as the "risk management division."

A volunteer department may have its liability insurance and member medical coverage through a private insurance company. Many companies throughout the country specialize in liability insurance and medical coverage for volunteer fire and rescue departments. These organizations have representatives who work with the department to administer the medical coverage and to identify ways to reduce liability through a safety and health program.

Although it is not a requirement, an organization is morally obligated to provide a safety program that includes the accident and injury record-keeping and documentation process.

Legal Liability

As the ISO, you should understand the record-keeping and documentation process to follow if an accident or injury occurs. In order to protect the department against liability and to protect personnel, you have to know the process.

Types of Documentation

Fire or EMS Department

Every organization must have procedures that thoroughly document accidents to department equipment and vehicles, or injuries to personnel. The insurance carrier will dictate what information and documentation are required. A department may have specific procedures (SOPs) or policies covering these requirements. Included in this process are specific guidelines that require a supervisor or the ISO to report serious injuries or accidents through the organizational chain of command. This notification process may include notifying the risk management division of a vehicle accident that could pose liability problems for the department or organization.

Municipal, County, or State Documentation

The governmental entity may have specific requirements for recording and documenting accidents and injuries, based on state or local laws. The department must be aware of these requirements. Failure to comply with them could expose the department to liability.

Types of Incidents to Be Documented

What types of incidents should be documented?

Accidents that result in damage to department vehicles or apparatus, damage to department equipment or facilities, damage to private or civilian property, and near misses to any of these must be documented. Lost or stolen equipment also would be placed in this category.

Injuries to personnel while on the job, regardless of the severity, must be documented (see Figure 4-3). Near misses should also be documented. An injury report must be filed for civilians injured while on department property.

Document incidents of personnel exposed to hazardous material. The medical effects of an exposure to a hazardous material may not appear immediately; when they do, treatment might be needed on a long-term basis. Without documentation, claims for medical treatment could be denied. The same is true for job-related exposure to a communicable disease. (These records must be maintained for 30 years after employment, per OSHA 29 CFR 1910.1030.)

If a firefighter or EMS provider dies in the line of duty, there are specific guidelines that must be followed to comply with the requirements of the Public Safety Officers' Benefits (PSOB) Act so that the public safety officer's family or dependents can receive the benefit. OSHA investigates job-related fatalities. Consider OSHA a resource in the event of an employee death.

FIGURE 4-3 All types of injuries to responders—no matter how serious or insignificant they may seem—should be documented. Photo by Louis Carter, Jr., District of Columbia Fire and EMS Department.

Typically, the injured employee's supervisor is responsible for investigating and completing preliminary accident or injury reports. This should be explained in detail in the departmental reporting procedures. Regardless of the severity of the incident, the supervisor must investigate and complete the documentation. If the incident goes beyond the capabilities and expertise of the immediate supervisor, the safety officer can assist in completing the investigation.

Safety Officer Responsibilities

Although the safety officer should be notified of all accidents and injuries, the severity of an accident or injury determines whether the safety officer will become involved in an investigation. Examples of incidents requiring the involvement of the safety officer include the following.

- An injury to a firefighter or EMT/paramedic that requires medical treatment at a hospital
- The admission of a firefighter or EMT/paramedic to a hospital because of a job-related injury or illness

- Department vehicle accidents
- Department vehicle accidents that involve injuries

Department policy should specify which incidents require safety officer involvement; however, the safety officer should review all investigation reports. This allows the safety officer to serve the department better through use of resources. Department personnel must notify the safety officer if an incident occurs.

Resources available to help the safety officer in an investigation include the following.

- Law enforcement personnel and laboratories
- Mechanics (for equipment failure)
- Structural engineers
- Industrial hygienists
- Fire protection engineers
- Testing laboratories
- Fire investigators

These individuals will be able to assist and provide valuable expertise during an investigation.

Reporting Information

At a minimum, the following information should be included in injury investigation reports.

- Date and time of the incident
- Name of the employee's immediate supervisor
- Exact street location of the incident
- Type of activity the member(s) was engaged in at the time of the incident
  - fireground
  - station duty
  - responding to or returning from an incident
  - performing department training
  - operating at nonfire emergencies (e.g., EMS incidents like vehicle accidents or special operations like confined-space incidents)
- Employee name and appropriate information (address, social security number, telephone number, etc.)
- Employee's current department assignment (i.e., firefighter, fire captain, fire inspector)
- Brief but thorough description of how the incident occurred
- Description of the type of injury or description of the damage to equipment or vehicle
- For lost or damaged equipment, the cost of repair or replacement
- Brief description of how to prevent the accident or injury from recurring

Fire departments using the National Fire Incident Reporting System (NFIRS), developed by USFA, have a "firefighter casualty report" that collects this and other information on incident-related injuries. It does not, however, apply to nonincident-related injuries, so an additional information collection system must be used.

Accident and Injury Analysis

The investigation process and procedures are critical to documenting a death, injury, or accident. The most important factor is to determine what happened. Unfortunately, this is often not how the investigation progresses. For the most part, investigations fix blame rather than find facts. The person in charge of an investigation often tries to point out who was at fault rather than to learn what truly happened.

The investigation should begin immediately and be conducted according to department protocol to ensure that complete and through information is gathered. The immediate supervisor analyzes and completes reports for the majority of the accidents and injuries that occur. The immediate supervisor is the most logical individual to study the situation. Paperwork is very important, because supervisors are required to document particular incidents; information in the report can indicate trends, patterns of types of injuries, or accidents.

For situations that require more detailed analysis, the supervisor can contact the safety officer for assistance. The safety officer becomes involved when serious accidents or injuries occur. If the situation is incident related, the safety officer may request assistance from the HSO or the infection control officer. These individuals are able to be involved in lengthy investigations, whereas the ISO may or may not be available.

The safety officer should know what resources can be accessed from inside and outside the department, depending on the severity of the incident. Once an incident occurs, it is too late to start trying to find agencies or individuals that have specific expertise.

Employee Fatalities or Serious Injuries

If an employee dies or is seriously injured in an occupational accident, outside agencies respond to conduct an investigation and provide an analysis. These agencies can include law enforcement for motor vehicle accidents, OSHA inspectors, and insurance investigators. Some information must be made available to meet the PSOB requirements. Investigators who represent employee interests may be part of this process, as well.

Confidentiality

When medical documentation is involved, it is imperative that patient confidentiality be maintained. The information required for accident, injury, and illness reporting should not contain medical treatment information or confidential medical diagnoses.

For example, medical treatment for illnesses associated with a communicable disease is confidential. The medical treatment and the outcome of the diagnosis are confidential information between the patient and physician. As a rule, departments should not maintain medical records, but if there is no alternative, the medical reports should be kept separate from personnel files. The only people having access to these files should be the chief and fire department physician.

Confidentiality is a critical issue when dealing with medical data, and it must be handled in a very sensitive manner. Witness statements taken at the scene of an accident or injury may be confidential information, especially if negligence is involved. If you are unsure of how to deal with a situation, contact the department's legal representative for assistance.

Post-Incident Analysis

After an incident, a department must systematically evaluate and review the operations that took place. This process is commonly known as a critique, post-incident critique, or post-incident analysis (PIA). All of the significant players involved in the incident should be included in the process. As depicted in Figure 4-4, an informal debriefing can immediately follow an incident, but it cannot take the place of a formal PIA for major incidents.

The PIA includes a basic evaluation of the circumstances and the operations. It also includes the effect of the circumstances and operations on the safety of the personnel present at the incident. NFPA 1500, Section 6-8.1 states, "The fire department shall establish requirements and standard operating procedures for a standardized post-incident analysis of significant incidents or those that involved serious injury or death to a firefighter."

It is very difficult for most of us to evaluate ourselves effectively when the operations and outcome of an incident do not go well. The PIA is the process necessary to support change in SOPs, policy, and/or responsibilities if the organization is to benefit from mistakes.

Safety and Health Issues

The PIA must include all safety issues or concerns that relate to the incident, both positive and negative. An ISO must present the significant events that affected the incident from a safety standpoint. If all operations worked well, with personnel operating as required and wearing proper PPE, and no accidents or injuries occurred, then the ISO should acknowledge these positive actions. The same is true for operations or actions that do not follow SOPs or department policy.

FIGURE 4-4 An informal debriefing (sometimes referred to as a "tailboard talk") can take place immediately following an incident, but a formal post-incident analysis is important for significant incidents. Photo courtesy Chesterfield County (VA) Fire and EMS.

Should a fatality occur, the PIA process will be significant. A preliminary report of what happened should be made as soon as possible to eliminate or rebut any rumors and to state exactly what occurred.

Personnel injuries or health exposures should be reviewed and discussed. Remember that this is not done to embarrass, belittle, or point fingers at a particular person or persons. The ISO must identify what happened and determine how the organization can learn from the incident.

Key issues for the ISO to discuss during every PIA, whether in a positive or constructive manner, include the following.

- PPE
- Accountability
- Health concerns

These particular issues should be covered from the standpoint of what the ISO identifies as correct or incorrect actions. This reinforces "good behavior" and constructively corrects "improper behavior." This consistent identification process also helps to evaluate overall department operations, not to identify personal issues. Most importantly, the ISO must maintain a positive attitude and be proactive throughout the PIA.

The PIA will identify operations and procedures to change or revise for the safety and welfare of department personnel. These changes or revisions will be in SOPs, policy, or training and education. Once issues have been identified, an action plan should be developed to make changes. The action plan should include what changes or revisions need to take place; who is responsible for making them; the dates the changes will be made and when they will become effective; and any other details of the action plan.

Summary

Record keeping and documentation are critical elements of a department's safety and health process. It is the ISO's responsibility to ensure that procedures are followed; documentation is completed in an accurate and timely manner; and all relevant information is included. Trends can be identified, and action can be taken to reduce hazards to fire and EMS personnel.

Proper documentation of ISO actions at emergency incidents and of any type of investigation conducted by the ISO or HSO is important from a liability standpoint, as well. Liability can have negative consequences for a department or organization in a variety of ways. The more a department does to manage liability potential, the more productive it becomes.

The PIA is a critical part of the health and safety process. During each PIA, time must be given to safety and health concerns. The ISO will be the individual to focus on issues that need to be addressed and corrected. Although this process may not be as exciting and stimulating as other ISO functions, it is one of the most important.

Part Two

Administrative Aspects—The Health and Safety Officer

5
Risk Management

Risk

Each of us faces risks every day. Every action that we take carries with it the chance that we might be killed or injured. Risk has an impact on us as we work and as we play; escaping risk is impossible. No activity is totally without risk.

In most cases, the risks that we take are very minor. The risk of being hit by lightning is very low, yet few of us are comfortable in an open field during an electrical storm. Likewise, the risk of crashing in a commercial airliner is minor, but it is always on everyone's mind during take-off and landing.

Insurance companies are very familiar with the concept of risk. Car insurance premiums for an 18-year-old single male are much higher than they are for the same man after he is married and 30 years old, provided he has not had any major losses. The cost of the insurance reflects the risk that is associated with the driving behavior of a person at different times in his/her life.

We each face risks in our lives at work and away from work every day. Nothing we do is completely without risk. Sometimes, we do not think about the risks that we face, but they are still there.

As each of us lives our life, we avoid risk. Sometimes, we think about avoiding it, and sometimes we just do it without thinking. If you choose not to enjoy the thrill of hang gliding because you feel it is too risky, you have completely avoided the risk. But, not every risk can be avoided completely. In other cases, we can avoid a significant risk but not be completely protected from it. If the law enforcement agencies in your area begin referring to a stretch of road as the "mile of death," you may choose to use another road during your travels. There still is some risk associated with driving, so you have not completely avoided the risk—just reduced it. Think about the risks faced by individuals involved in the following situations, and how they can be reduced.

- Driving a car
- Riding on a bus
- Spending a night in a high-rise hotel
- Bungee jumping
- Attacking a car fire

- Treating the injuries of someone who was trapped in a vehicle collision
- Performing an interior attack on a structural fire

Because of the significant risks faced by emergency responders, attempts are made to identify and control the risks that we face as we do our jobs. This activity is known as *risk management.*

Risk Management

Risk management is the method used to reduce exposure to risks. As stated earlier, many risks cannot be completely avoided. Fire fighting, emergency medical care, and special operations are extremely hazardous activities.

Risk management is a tool used in every industry to control risks. The best way to deal with any specific risk is to avoid it completely. In many situations, this is not possible, so other strategies for dealing with it must be developed.

Risk management for emergency response organizations is divided into three categories: nonemergency risk management, pre-emergency risk management, and emergency incident risk management.

Nonemergency risk management looks at the hazards common to all workplaces. This type of risk management might include fire inspections of fire stations and the management offices of emergency response organizations. The risks encountered in these places are no less deadly than those encountered on the emergency scene, but their frequency and severity are lower than those risks faced on the emergency scene.

Pre-emergency risk management activities take place prior to the emergency. These activities can have a major impact on the safety of members working at the scene of an emergency. Some examples of pre-emergency risk management activities include selection of PPE used by a department; effective SOPs and training; warning devices and reflective striping on apparatus; and pre-emergency planning of target hazards.

Emergency incident risk management is the method used to reduce the risks faced by responders at the scene of emergencies and, in many cases, training evolutions. It is primarily the responsibility of the ISO. Emergency incident risk management is covered more completely in chapter 8.

Classic Risk Management Process

The classic risk management process—used prior to, during, or after the emergency—is divided into five steps. Most often, it is used prior to the emergency.

IDENTIFY RISKS

Risk identification is the process of making a list of things that might go wrong with an operation. A good rule of thumb is to anticipate the worst that can happen when identifying risks. If plans are formulated for a worst-case scenario, anything less can be handled. Sources for this information may include past department accident and injury statistics, input from members of the department, and knowledge of the experiences of other emergency service providers.

EVALUATE RISKS

Once the risks are identified, they can be evaluated both from a frequency and severity standpoint. Frequency is the likelihood that a risk will be faced. Some risks are present at every emergency, and some may be faced only once every 2 years. Severity is an indication of how much damage or injury can be caused by the risk. Risks do not go away by themselves. If your agency has experienced an injury or a death in a particular situation and no means to manage the risk have been put into place, your agency runs the risk of experiencing the same situation again.

By using the information gathered in the identification step, you can classify risks by severity.

ESTABLISH PRIORITIES

When considered together, frequency and severity information will help establish priorities for action. Any risk that has a high likelihood of happening (frequency) and has a great potential for damage and injury (severity) should be handled first. On the other hand, risks that have a low likelihood of happening and have a low potential for damage and injury would be lower on the priority list.

A good tool for the risk classification step is shown in Figure 5-1. Each risk should be placed in one of the four areas. For example, a severe risk that is faced all of the time would appear in the upper-right box; a risk that is not severe and is faced on an infrequent basis would be placed in the lower-left box. In general, high frequency, high severity risks would be addressed first, and low frequency, low severity risks would be addressed later.

The prioritization process is not a simple task. There is no one correct priority for each risk. The decision concerning which risk to handle first depends upon

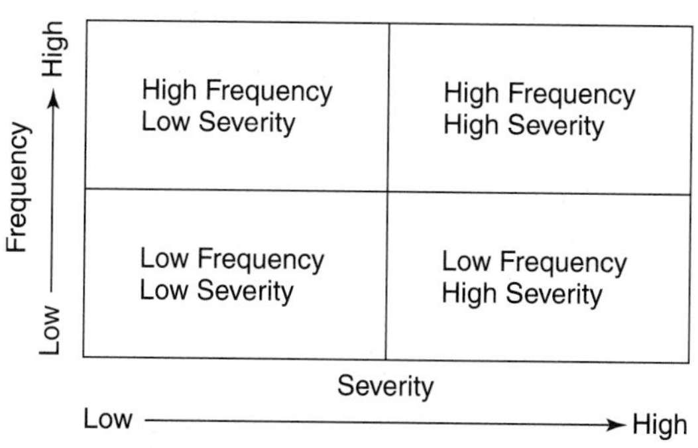

FIGURE 5-1 Risk Prioritization Matrix.

local factors, such as the availability of resources, the ease of addressing each risk, and the time involved in addressing each risk.

CONTROL RISKS

After all of the risks facing an operation have been identified, it is time to attempt to find solutions to each risk. Some risks, especially those that occur during emergencies, cannot be completely controlled, but the severity of the risk can be addressed. For example, the fire department cannot control the strength of a burning structure. However, the department can control the severity of the risk by restricting access to the area where the structure will fall and by implementing other standard operating procedures that dictate when it is permissible to enter a burning building.

Several methods are used to control risks. They are risk avoidance, risk control, and risk transfer.

RISK AVOIDANCE

This method of risk management does little to protect emergency responders because it involves the complete avoidance of a particular risk. If a bridge in a community is hazardous, use of the risk avoidance method would not allow fire apparatus or other emergency vehicles on the bridge. Thus, the risk from the bridge is completely avoided. The safety of the alternate routes would need to be a consideration.

This method is of little use in emergency operations. Emergency scene operations are dangerous, but they cannot be completely avoided. If emergency responders avoided the emergency, who would provide the medical and fire control services needed at the scene?

RISK CONTROL

This is the most visible type of risk management. Control measures include personal protective equipment (PPE) and the use of SOPs, such as an incident management system. The risks to emergency responders are identified before the emergency, and control measures that will have an impact on the safety of responders when the emergency occurs are put in place.

A collision between emergency response vehicles and civilian vehicles at intersections is a risk that has high severity and high frequency in the emergency services. The development of SOPs that require a full stop at red lights and other negative right-of-ways situations can help to control the severity and the frequency of this risk.

RISK TRANSFER

This method involves transferring a risk to another organization or, in the case of a financial risk only, to insurance. If an emergency response organization concludes that inspections of gasoline tankers are just too risky to be performed by their members, the organization can contract with an outside agency to perform the inspections. Thus, the risk is transferred from the emergency organization to an outside organization.

In cases where the risk is purely financial, such as risk of a fire in an unoccupied building owned by an emergency response organization, insurance can be purchased that protects the emergency response organization from the risk. It should be noted that risk transfer does not eliminate the risk or reduce the chances of a fire; it only provides compensation if something goes wrong.

MONITOR RISKS

Once control measures are in place, the effectiveness of the control measures must be monitored. In the case of an SOP that requires full stops at intersections, if emergency vehicles continue to be involved in collisions, the SOP and the success of its implementation would need to be addressed.

Fire Department Risk Management

NFPA 1500, *Standards on Fire Department Occupational Safety and Health Program,* requires that all fire departments develop and adopt an official written risk management plan. The plan must cover all fire department facilities and operations and use the classic risk management methods described previously. Risk management plans also should be developed for other emergency response organizations, such as EMS departments.

The risk management plan should look at every function performed by the organization, nonemergency and emergency. The plan should identify risks associated with each operation; evaluate and classify the risks faced in each operation by frequency and severity; prioritize the risks associated with every operation and the risks faced by the organization as a whole; and suggest the development of risk control measures. In addition, the HSO should monitor the effectiveness of the risk control measures that are used. Excerpts from a fire department risk management plan are shown in Figure 5-2.

The HSO as Risk Manager

The HSO should be involved in all operations of the emergency response agency. The HSO brings a safety perspective to all meetings and can provide valuable input to other members of the agency concerning the safety of responders.

The HSO is the risk manager for the fire department or EMS department. When the HSO is involved in the specification of apparatus or equipment, he or she is really acting as an internal safety consultant. The HSO may not be involved, for example, in selecting the components of the drivetrain of a vehicle, but it is

OPERATION AND RISK ID	FREQUENCY/ SEVERITY	RANK/ PRIORITY	CONTROL MEASURES
		Emergency-Mode Driving	
Collisions with other vehicles	Low/High	High	Emergency-mode driving regulations Traffic control devices Warning lights/sirens Speed limitations Public education campaign Not all calls require emergency mode
Collisions with objects	Low/High	High	Emergency-mode driving regulations Use of backers Automatic door openers
Injury to passengers	Low/High	High	Seatbelt provision and use Enclosed equipment in cab
Injury to others	Low/High	High	Public education campaign Warning lights/sirens Traffic control devices

FIGURE 5-2 Excerpts from a Fire Department Risk Management Plan.

appropriate for the HSO to recommend disc brakes, antilock brakes, and a supplemental breaking device like a magnetic driveline retarder.

Hazards of Operations and Equipment

Every operation carries with it some risk to responders. The use of every piece of equipment carries with it some hazard to the user. These risks can be minimized by their early recognition and through the use of control measures, such as the required use of safety equipment and the use of SOPs.

A listing of hazards and control measures by subject follows.

Apparatus and Vehicles

- All personnel are seated in enclosed areas.
- Seatbelts are provided for all riders.
- Equipment (including SCBA) inside the passenger compartment is secured or enclosed.
- SCBAs stored outside the passenger compartment are secured in a compartment or enclosed container.

- Driver safety training is provided to all drivers.
  - Stopping distances
  - Emergency maneuvering
  - Aerial operations
  - Pump operations
  - Backing hazards and the use of spotters
  - Vehicle does not move until all passengers are seated/belted
- Vehicle visibility in emergency and nonemergency models.
- Emergency mode driving procedures.
  - Stop signs and red lights
  - Negative right-of-way situations
  - One-way streets
  - Speed limits
  - Challenge and response routine between the driver and the officer
  - Railroad crossings
  - The dangers posed by engine brakes, retarders, and brake limiting switches
- Weight and brake capacity.
  - Loading of new vehicles
  - Weight balancing, side to side
  - Tank baffling
  - Military truck conversions, fuel tanker conversions
  - Antilock brakes on newer chassis
  - Automatic chains for colder climates
  - Supplemental braking devices such as jake brakes, transmission retarders, and magnetic driveline retarders
- Passenger heat and noise reduction. Sirens, radios, air horns, engine, etc.
- Patient lifting height for ambulances.
- Child safety seats for pediatric riders.
- Restraints for EMS providers and others in the back of an ambulance.
- Compliance with the appropriate federal or NFPA standard.
- Preventive maintenance and inspection program.

Facilities

- Smoke detectors and automatic sprinklers.
- Diesel exhaust emissions control (see Figure 5-3).
- Carbon monoxide detectors.
- Standard fire safety measures regarding storage, electrical appliances, safety with cigarettes and open flame, and regular inspections.
- Dedicated area for the cleaning of EMS equipment, contaminated PPE, etc. (This should never be brought into the living area of the station.)
- Dedicated area or container for disposal of medical or biohazardous waste.

FIGURE 5-3 Many fire EMS departments have installed diesel exhaust extraction systems to minimize responders' exposure to the potentially harmful effects of diesel exhaust. Photo by Gordon M. Sachs.

Protective Clothing and Equipment

- Compliance with the appropriate NFPA standard.
- *Full* structural protective clothing: helmet, hood, coat, gloves, SCBA, flashlight, trousers, and boots.
- Protective clothing for EMS operations: exam gloves, eye and face protection, drapes, gowns, and suits.
- Clothing and equipment—and a means for cleaning them—must be provided by the agency.
- Fire-resistant uniforms.
- Clothing worn by volunteer firefighters under structural protective clothing.
- SOPs that govern the use and maintenance of protective clothing and equipment.
- The right protective clothing for the occasion: structural, wildland, proximity, hazmat, etc.

Equipment

- Proper training for all members expected to use the equipment is a basic requirement.

- Weight, center of gravity, and ease of carrying. Are handles in places that make the equipment easy to carry (portable pumps)?
- Consideration of noise levels produced by the equipment.

Fire Fighting

- Use of an incident management system.
- The use of ISOs on the scene of an emergency.
- Working in teams of at least two members in the hazard area.
- At least two members outside the hazardous area, capable of firefighter rescue, whenever firefighters are operating inside the hazardous area.
- The use of risk management techniques on the scene of the emergency. Risk versus benefit.

Emergency Medical Services

- Use of proper levels of communicable disease protective equipment and clothing.
- Proper decontamination of equipment.
- Scene control. (Traffic accidents, violent incidents, etc.)
- The use of an incident management system.
- Proper lifting techniques.

Training Exercises

Some of the most dangerous nonemergency activities that responders can perform are live-fire training and training for special operations, such as hazardous materials emergencies or high-angle rescue. Risks at these training exercises mirror the risks in emergency operations (see Figure 5-4). The HSO should apply the same risk management techniques used at emergencies to these events.

NFPA 1403, *Standard on Live-Fire Training Evolutions in Structures,* provides safety and operational guidance for live structural fire training and should apply the same risk management techniques used at emergencies to these events.

In his or her role as the agency's internal safety consultant, the HSO should be involved in planning for all training exercises. A safety action plan should be developed by the HSO for each exercise and reviewed with the IC of the exercise.

The safety action plan should take the following into account.

- Type of exercise to be conducted.
- Risk management evaluation of the hazards that may be encountered.
- Review of any applicable standards or regulations governing this type of exercise.
- Preparation of a checklist of hazards to be aware of during the exercise.
- Speaking points for the safety section of a pre-exercise briefing of the participants.

FIGURE 5-4 Live-fire training evolutions can be just as dangerous as the "real thing." Risks at these training evolutions mirror the risks in emergency operations, and the same risk management techniques should be applied. Photo by G. Emas, Edmonton (Alberta) Emergency Response Department.

Extremely Hazardous Operations

Some types of operations pose an extreme hazard to responders. Special attention by the HSO is warranted.

Technical Rescue and Hazardous Materials Operations

Almost every emergency response agency responds to technical rescue or hazardous materials calls each year. Technical rescue calls may involve trench collapses, water rescue, ice rescue, rescue from heights, or the rescue of persons trapped by a building collapse. Hazmat calls include large flammable liquid fires, gas leaks, and other incidents involving chemicals. Although every agency faces these risks regularly, their frequency is much lower than most fire or EMS operations.

Firefighters and EMS workers are task oriented and tend to make do with the tools at hand: Their natural inclination is to help those in need, regardless of the personal risk. Technical rescue and hazmat situations usually require specialized equipment and training, as shown in Figure 5-5. The agency risk management plan should identify a role for first responders to control access to the scene and call in technical experts.

FIGURE 5-5 Hazardous materials and technical rescue incidents can be extremely dangerous. The department's risk management plan should address the roles to be performed by responders, as well as those performed by technical experts. The HSO should consult technical experts when developing the plan and SOPs for emergency response to these types of incidents. Photo by G. Emas, Edmonton (Alberta) Emergency Response Department.

The HSO does not need to be an expert on all of these hazards. However, a general understanding of the risks and the methods used to control these incidents is necessary. The HSO may choose to consult with experts from within or outside of the agency when developing SOPs for these types of incidents. These types of operations also tend to be heavily regulated on the local, state, and national level, and these concerns should be addressed in the SOP.

Incidents Involving Violence

A much more common risk faced by responders today is the response to an incident that involves violence. In the past, fire and EMS workers were considered neutral in these situations. The overall positive feeling of the community toward firefighters and EMS workers provided a feeling of security. Changes in our society and the increasing role of the fire service in EMS have increased our exposure to these risks.

Some members of the community may view firefighters and EMS workers as extensions of the government. This point was brought home to all firefighters and EMS workers during the civil disturbance of 1992 in Los Angeles and the surrounding area. Firefighters and EMS workers required police or National Guard escorts in order to perform their jobs, and one firefighter was seriously injured by gunfire.

Our exposure to violence can come in many forms, some much less spectacular than civil disturbance. Firefighters and EMS workers are exposed to violence when responding to bomb threats, assaults, shootings or stabbings, domestic violence, robberies, hostage situations, or drive-by shootings. Violence may also occur when firefighters and EMS workers support police operations, such as raids on drug houses or drug labs.

There are at least two ways to address our increased exposure to these hazards: through community involvement and through the use of SOPs. The neighborhood fire station has traditionally been a gathering place for the people of the community. Many volunteer fire stations have banquet and meeting facilities attached to their stations. These facilities provide income to the department and an opportunity for the members of the department to keep in contact with the residents of their community. The downtown fire station with the apparatus doors rolled up and firefighters outside watching the world walk by is a thing of the past in many communities. Closed doors and barbed wire around the parking lot have replaced the open station. Many firefighters do not even live in the community they serve.

What this all adds up to is a disassociation of the local firefighters from the people that they serve. This insulation from the community through closed doors or lack of contact degrades the traditional bond that we have enjoyed with our customers. One way to prevent violence toward firefighters and EMS workers is to keep involved with the people in your first-due area: Open the station to visitors, and don't let the actions of a few generate a siege mentality. A caring and respectful attitude shown toward the people you serve will go a long way.

When violence does erupt, firefighters and EMS workers have no place in the line of fire. We do not have the equipment or training to survive these situations. Firefighters and EMS workers should stay out of the hazard zone until law enforcement officials have stabilized the situation.

If a person is in need of medical care as the result of a violent attack, law enforcement should secure the area to allow firefighters and EMS workers to enter, or they should move the victim to a safe area so that treatment can be initiated. These options should be addressed in the agency's SOP.

If firefighters or EMS workers are directed by their SOP to stage away from all incidents involving violence, other problems can arise. The person in the best position to judge the safety of the scene is the company officer, in communication with law enforcement.

Firefighters and EMS personnel should wear uniforms that differ in appearance from those of law enforcement personnel to avoid any confusion about an individual's role at the scene.

Summary

Everyone faces risks every day. It is unavoidable. In the emergency services, some of the tasks we perform involve significant or even extreme risk. To deal with these risks and lessen the possibility of death or injury, the concept of risk management is used. Nonemergency risk management deals with day-to-day hazards that might be faced in any workplace. Pre-emergency risk management involves taking proactive steps to minimize the risks responders will face while responding to or operating at an emergency incident. Emergency risk management involves reactive steps that are taken to minimize the risks associated with operations at an emergency incident.

Classic risk management is utilized in the nonemergency and pre-emergency risk management processes. This involves five steps—identify risks, evaluate the risks, establish priorities, control the risks, and monitor the risks. Emergency risk management is too dynamic and fast-paced to utilize classic risk management. Instead, decisions are based on training, experience, and intuition, utilizing the emergency incident risk management guidelines as a framework for the decisions. These guidelines are:

Risk a lot only to save a lot;
Risk only a little to save a little; and
Risk nothing to save what is already lost.

NFPA 1500 requires that each fire department have a written risk management plan. Typically, the HSO develops and manages this plan. The plan should address apparatus and vehicle hazards, facilities, personal protective equipment, and all types of operations performed by the department. These operations include emergency operations, training exercises, and other hazardous or potentially hazardous operations.

6

Health Maintenance

For more than 10 years, the leading cause of firefighter death on the fireground has been heart attacks. Approximately one-fourth of the firefighters who died of heart attacks were younger than 40 years old. The majority of those were not exerting themselves physically at the time of the heart attack. The leading cause of firefighter deaths overall (fireground and nonfireground) is heart disease and stress.

A passage in the USFA publication, *Firefighter Fatalities in the United States, 1998,* clearly describes the importance of health maintenance for emergency response personnel:

> From 1995 through 1998, 171 firefighters have died of heart attacks and strokes. Many of the firefighters who died had pre-existing medical conditions that placed them at higher risk for heart disease and strokes. In many cases, these conditions were known to the fire department. In many more cases, the health of the firefighter was unknown prior to their death.
>
> In 1998, at least 19 of the 41 firefighters who died of heart attacks had pre-existing conditions such as prior heart problems, bypass surgery, or diabetes, which can indicate the possibility of heart and cardiovascular disease.
>
> *Firefighter Fatalities in the United States, 1998,* USFA, p. 35

From 1989 to 1998, there was an average of over 95,000 injuries each year, of which 50 percent or more occurred on the fireground. The most common types of fireground injuries were strains, sprains, and muscular pain, followed by wounds, cuts, bleeding, and bruises. Burns accounted for almost 10 percent of these injuries, with smoke or gas inhalation responsible for almost another 10 percent. The top causes of fireground injuries were overexertion and falls, slips, and jumps.

Although national death and injury statistics for nonfire EMS providers are not available, one can infer that the results would be similar to firefighter death and injury statistics in many ways. For example, the number of injuries at nonfire emergencies during the period 1989–1998 averaged over 14,300 each year, or about 15 percent of the total number of firefighter injuries (see Figure 6-1).

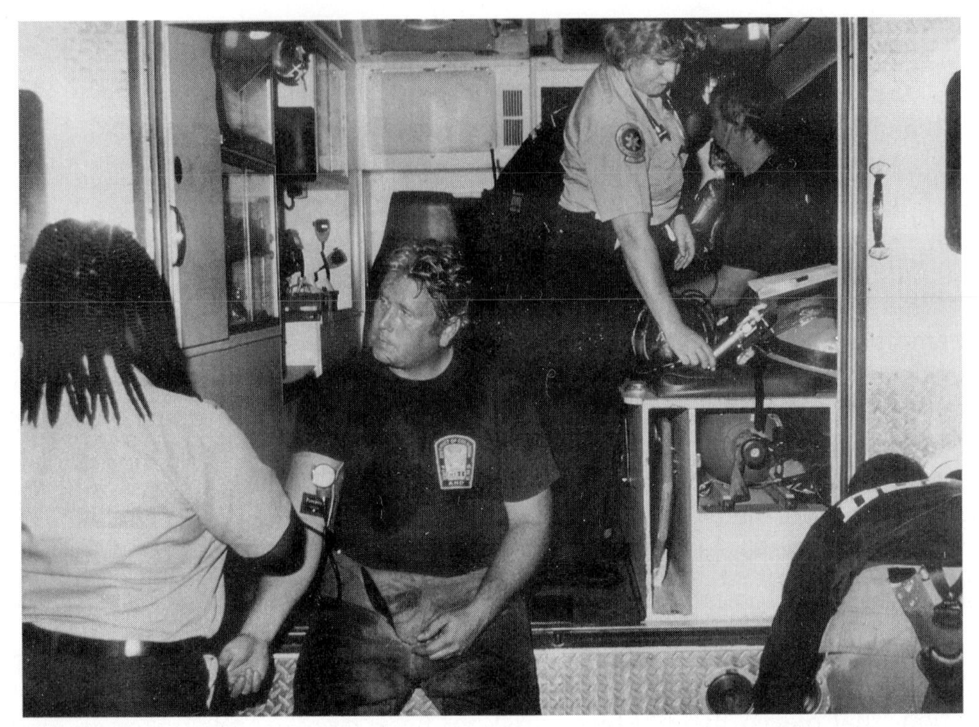

FIGURE 6-1 The number of firefighters injured each year is staggering. The number of EMS responders injuries is also likely to be very high, but such statistics are not available. Photo by Louis Carter, Jr., District of Columbia Fire and EMS Department.

High-Risk Factors for Firefighters During Fireground Operations

The following risk factors can affect firefighters operating at emergency scenes and increase the likelihood that injuries will occur.

Lack of Physical Fitness

Overweight firefighters are at an increased risk for many common fireground injuries such as strains and sprains. In addition, being overweight can be a factor in the development of cardiovascular disease.

A lack of regular physical exercise contributes to the problem of being overweight and has a general effect on cardiovascular fitness and muscle strength. Both of these areas can affect firefighters' well-being when firefighters must perform strenuous tasks at an emergency scene.

History of Illnesses

Alcoholic firefighters have more chronic health problems than do other firefighters. They face an increased risk of heart problems and strokes. Nearly all of the body's tissues and organs are affected by alcohol abuse.

Most firefighters are unaware of any cardiovascular problem until they develop symptoms or have acute problems. Warning signs such as high blood pressure or a family history of heart disease should create an awareness of the possibility that problems could develop at an emergency incident.

Firefighters who return to work after an illness are not always in good physical condition. If they are physically stressed before they have recovered completely, they may require more frequent breaks and become tired more easily.

Diarrhea can cause an electrolyte imbalance. When firefighters are involved in strenuous activities, these electrolytes can become even more imbalanced. When levels of important ions like potassium and sodium are disturbed, smooth muscles do not contract as well, especially the heart.

Medications

Certain medications can have a long-term effect on the health and well-being of an emergency responder. Some types of common, over-the-counter medications that can affect a responder's ability to operate safely or effectively include the following:

- Antihistamines—Benadryl, Actifed; normally taken for colds and sinus problems
- Diuretics or hypertensive medications—Lasix, Inderal, Isoptin, Procardia, Cardizem; medications prescribed for high blood pressure or other heart-related problems
- Stimulants—caffeine, decongestants, or diet pills

Smoking

Several thousand firefighters die each year from the effects of smoking. Smoking causes more lung cancer, heart disease, and lung disease than any other factor. Smoking increases the amount of carbon monoxide carried in the blood. This reduces the amount of aerobic capacity firefighters have to perform the job.

Hearing Loss

Noise can strain the inner ear and cause a temporary hearing loss. If the ear has time to rest, hearing is restored. If the exposure to noise is continual, the ear can lose its ability to recuperate, and permanent damage can occur. Permanent hearing loss is caused by the destruction of cells in the inner ear.

Fire departments should consider implementing a program to protect firefighters from hearing loss. Current OSHA noise requirements set a maximum permissible noise level (PEL) of 90 dB for an 8-hour period. Hearing loss as a result of noise exposure is recognized as a significant health hazard in the fire service. Hearing protection should be mandatory for all personnel riding on fire apparatus.

New firefighters should have baseline audiometric testing performed. All personnel should receive periodic testing. Recommended standards are included in NFPA 1582, *Standard on Medical Requirements for Fire Fighters and Information for Fire Department Physicians.*

A reference, *Hearing Conservation Program Manual,* is available from the USFA. This manual provides information on hearing loss and programs to help conserve hearing in firefighters.

Back Injuries

Back injuries account for the biggest category of workers' compensation and lost-time injuries in the workplace. The major cause of back injuries is improper lifting techniques. The key to avoiding back injuries is prevention. In addition to training employees on proper lifting techniques, employees should exercise regularly to build flexibility and strength.

Cancer

Many firefighters have developed cancer during their career or shortly after retirement. Although many types of cancer have been found, skin cancer is among the most common forms of cancer identified. It is important to follow medical examination schedules and address potential problems early.

Firefighter Wellness Programs

When most people are asked to define a wellness program, they usually describe a physical fitness or exercise program. These are only part of an overall wellness program. Defining a true comprehensive wellness program has been a topic of debate in the emergency services for several years. This debate culminated with a cooperative effort between the International Association of Fire Fighters and the International Association of Fire Chiefs in the establishment of a joint Wellness/Fitness Initiative. Working with fire departments across the country, the Fire Service Joint Labor-Management Wellness/Fitness Task Force has developed a medical wellness/fitness program, as well as other programs to improve firefighter health, wellness, and fitness.

A wellness program should be viewed as a pie. Fire departments often have one or two components of the program and call them a wellness program. There actually are several major components of a wellness program.

- Annual medical evaluation and periodic medical examinations
- Infection control
- Employee or member assistance program (EAP or MAP), including substance abuse
- Physical fitness program
- Emergency incident rehabilitation

Annual Medical Evaluations and Periodic Medical Examinations

Medical evaluations should occur both when new personnel enter the fire department and on an annual basis. A baseline measure of each person's physical condition and vital signs must be established, and a medical history of all personnel who may be involved in fireground activities should be developed. A physician does not necessarily need to conduct the *evaluation;* however, a physician should review the results.

A medical *examination* by a physician should occur when new personnel enter the department and periodically thereafter. NFPA 1582 identifies a schedule for medical exams based on the age of the firefighter. A baseline medical exam for firefighters might include the following tests or examinations.

- A basic medical exam by a licensed physician
- Electrocardiogram (EKG)
- Height
- Weight
- Blood pressure
- Heart rate (pulse)
- Respiration
- Complete medical history of illnesses/injuries
- Cholesterol level
- Triglycerides (fat level)
- Chest X-ray
- TB skin test
- Check for skin cancer
- Complete blood count
- Chemistry 23 blood test
- Hepatitis antibodies status
- Urinalysis
- Tetanus update
- Rectal exam for enlarged prostate or blood in stool
- PSA blood test for persons over the age of 50
- Carboxyhemoglobin (baseline CO level)
- Vision test
- Hearing test
- Current list of medications

References for medical evaluations and medical examinations can be found in NFPA 1500 (chapter 8, "Medical and Physical") and in NFPA 1582, *Standards on Medical Requirements for Fire Fighters and Information for Fire Department Physicians.*

NFPA 1500 states

8-1.1 "Candidates shall be medically evaluated and certified by the fire department physician. Medical evaluations shall take into account the risks and functions associated with the individual's duties and responsibilities."

8-1.2 "Candidates and members who will engage in fire suppression shall meet the medical requirements specified in NFPA 1582, *Standard on Medical Requirements for Fire Fighters,* prior to being medically certified for duty by the fire department physician."

8-1.3 "All members who engage in fire suppression shall be medically evaluated periodically as specified in NFPA 1582, *Standard on Medical Requirements for Fire Fighters and Information for Fire Department Physicians,* on at least an annual basis, and before being reassigned to emergency duties after debilitating illness or injuries. Members who have not met the medical evaluation requirements shall not be permitted to engage in fire suppression. . . ."

Infection Control

Firefighters often come into contract with individuals who have infectious or contagious diseases. The components of a complete infection control program are as follows.

- Exposure control plan for members at risk
- Training and education
- Engineering and work-control practices
- Hepatitis B vaccination
- Medical treatment, post-exposure evaluation, and follow-up
- Record keeping

VACCINATIONS

A hepatitis B vaccination is recommended for anyone whose job puts them in contact with blood and body fluids. The risk of contracting hepatitis B is far greater than that for other serious diseases. The vaccination does not always take effect with the first series, so antibody checks should be done. Recommended booster shots should be given at least every 5 years.

Hepatitis A vaccinations are useful in situations where exposures to floodwaters could occur. The symptoms of hepatitis A mimic those of the flu.

Tetanus shots are given to prevent lockjaw and usually are effective for 5 years. However, if an injury occurs that results in an open cut or wound and it has been 5 years or more since the last shot, another is recommended.

Every fall it seems that a new strain of influenza finds its way to the United States. Researchers try to identify the particular strain ahead of time and develop a vaccine that can lessen flu symptoms. Often, the cost of giving annual flu shots to an entire department is less than the cost of lost time by workers who contract the flu.

In some parts of the county, there have been recent outbreaks of measles. Some people have had to be revaccinated, depending on when they received their initial vaccination. Because EMS response could involve contacts with persons who have measles, fire and EMS departments should consider vaccinating personnel who are at risk of contracting the disease.

Vaccines for other illnesses and diseases are available. Fire and EMS departments should ensure that all personnel are vaccinated against any vaccine-preventable disease they may be exposed to.

POST-EXPOSURE EXAMS

Persons exposed to different hazardous materials should be given an initial medical examination to determine the presence of any dangerous chemicals. Because some of these chemicals may not be visible immediately, a routine follow-up medical check may be necessary a few months after the incident; follow-up checks could be needed for several years.

Checks for exposures to blood or body fluids should follow the procedures outlined in the fire department infection control plan. In some cases, a medical exam will be required after an exposure of this nature as well as tests for infectious or communicable diseases.

Employee Assistance Program

An employee assistance program (EAP), sometimes referred to as a member assistance program (MAP), is used to help employees who are experiencing personal problems. In some cases, these problems affect the employee's work. Most programs of this type are intended for the employee's family, as well. The goal of an EAP is to help an individual to work through or resolve a problem or series of problems and to help them to have a productive life.

If these problems have affected the employee's work performance, an EAP can help to rehabilitate and return the employee to work, rather than have the employee continue to the point where punishment or discipline is necessary.

The major components of an EAP are as follows:

- Substance abuse program
  - alcoholism
  - tobacco
  - drug addiction to legal or illegal drugs
- Stress management; critical incident stress management (CISM)
- Family relations

- Legal and financial concerns
- Health promotion

Guidelines for these programs are included in NFPA 1500.

Physical Fitness Program

Being physically fit can reduce the number of firefighter injuries and deaths, but physical fitness programs must be developed that are comprehensive and aimed at improving overall firefighter health. The fire components of a complete physical fitness program are as follows.

- Medical screening
- Fitness assessment
- Fitness standards
- Exercise program
- Nutrition

Medical Screening

Firefighter medical examinations should be done according to the guidelines in NFPA 1582. A baseline medical examination is performed when an employee enters the department. An annual evaluation is performed to determine whether a member is able to perform job functions.

Fitness Assessment

Assessments of individual levels of fitness are based on the following.

- Cardiovascular fitness (aerobic)
- Muscular strength
- Muscular endurance
- Flexibility

From the results of these tests, a qualified fitness coordinator can develop an individual exercise program.

Fitness Standard

A much-debated topic in the fire service concerns an acceptable fitness standard. NFPA 1500 requires that a standard be defined. Tests to determine firefighter fitness must be valid. The IAFF, in its manual *Developing Fire Service Occupational Health Programs,* lists three types of validity tests:

- Content validity—test elements are similar or identical to those of the job being tested.
- Criterion validity—uses statistical tests to predict job performance.

- Construct validity—measures underlying theoretical concepts. These tests are developed by experts and may not always measure the aspects of fire-fighting correctly.

What happens when a firefighter cannot meet the standard established by the fire department is a question that needs to be answered before a standard is developed. Maintaining a positive rather than a negative approach toward rehabilitation of the worker will prove to be more beneficial to the fire department and its members in the long run.

EXERCISE PROGRAM

An exercise program can provide tangible health benefits and reduce the chances of developing heart disease and some types of cancer. Regular exercise can help control weight and slow the aging process. An exercise program can also contribute to the mental health and well-being of individuals who participate on a regular basis.

NUTRITION PROGRAM

Included in a program of this nature are basic concepts of nutrition and dietary guidelines. What you eat affects weight control and other medical conditions such as high blood pressure and high cholesterol, both of which can lead to heart disease.

Emergency Incident Rehabilitation

Medical evaluations done in rehab can determine whether any emergency response personnel are in danger of collapse from cardiovascular complications. Many times, heart attacks begin with warning signs that are easily detected by the medical personnel who staff the rehab sector. In addition to chest pains, shortness of breath, and poor color, there are subtle changes in the vital signs that can warn medical personnel that a firefighter is not ready to return to firefighting duties. Increased blood pressure, irregular heart rate, disorientation, and a poor pulse oximetry reading indicate that a firefighter may need to rest and receive further evaluation before returning to duty.

A complete set of vital signs (blood pressure, pulse, respiration, and temperature) must be taken when a responder enters the rehab area. Some departments may use pulse oximetry to determine the saturation of oxygen in the blood.

Firefighters should have a place to rest and cool down for a minimum of 15 to 20 minutes; they should also receive refreshment (primarily water). Figure 6-2 shows an example of firefighters in a rehab area. This rehab area should be free from the stress of the immediate incident site and away from vehicle exhaust. In hot weather, it should be in the shade and have air movement (either from the breeze or a fan). After the rest period, it is important to record the vital signs again to determine whether any significant changes have occurred.

FIGURE 6-2 Emergency incident rehabilitation is an important facet of operations, because it can reduce the likelihood of responders succumbing to overexertion or stress—two of the leading causes of firefighter death and injury. Photo by Gerry Suftko, Mesa (AZ) Fire Department.

Anyone having abnormal vital signs upon arrival in rehab should be checked more frequently. Further monitoring may be needed. In cases where life-threatening vital signs are found or when a firefighter complains of chest pain or shortness of breath, they should be transported to the hospital for further evaluation.

The rehab area is not the ideal place to treat firefighter injuries or illnesses. Try to have any personnel who require treatment receive it at a separate treatment area staffed by dedicated EMS personnel, as shown in Figure 6-3. In some cases, it may be easier to move a rehab area than to move an injured/ill firefighter to a treatment area.

It is important to document all rehab activities. All personnel on the scene, including the incident commander (IC) and anyone in charge of a sector, division, group, or branch, should be checked at least once during an incident. Just because they have not been involved in the physical aspects of the incident does not mean that they are not under stress. There have been documented cases of ICs and other division, group, or sector officers who have had abnormal vital signs when they

FIGURE 6-3 All personnel at a major incident, including the command staff, should rotate through rehab. Members who require medical treatment should be evaluated by EMS personnel at a separate treatment area if possible. Photo by Gerry Suftko, Mesa (AZ) Fire Department.

were checked in rehab and who required further evaluation. Stress is not always caused by physical exertion. Two objectives of rehab are

- to provide an on-scene screening process to help determine if any personnel operating there are in danger of collapsing because of cardiovascular complications; and
- to monitor how firefighters are reacting physiologically and emotionally to the stress of the particular operation.

In July 1992, the USFA published *Emergency Incident Rehabilitation,* which presented the first actual guidelines departments could use to establish a rehab area at their emergency incident scenes. This 10-page manual includes information on responsibilities, establishment of the rehab area, guidelines, indices on heat stress and wind chill, and sample forms fire departments use. This manual is an excellent source of material for any fire department interested in information on rehab. It is included in its entirety in appendix D in this book.

Record Keeping and Documentation

NFPA 1500, Section 8-4, "Confidential Health Data Base," includes recommended guidelines for employee medical records. Section 8-4.1 states, "The fire department shall ensure that a confidential permanent health file is established and maintained on each individual member. The individual health file shall record the results of regular medical evaluations and physical performance tests, any occupational illness or injuries, and any events that expose the individual to known or suspected hazardous materials, toxic products, or contagious diseases."

Employee Medical Records File

The Occupational Safety and Health Act of 1970 requires employers to keep permanent records on employee exposures to certain potentially toxic or harmful physical agents; this is regulated in Part 1910, subpart Z. All employers are required to keep records of occupational injuries and illnesses. Part 1910, subpart C, deals with preservation of, and access to, these records.

Examples of reports to keep in an employee's medical record include the following.

- Annual physical report
- Return-to-duty reports
- Workers' compensation reports
- Records of vaccinations
- Exposure reports (hazmat, infectious disease)

Medical records are confidential and cannot be disclosed or released without an employee's written consent. They should be available to the employee/member or anyone who has the employee's written consent. An employee's medical record is to be maintained for the duration of employment plus 30 years.

Department Records—Rehab Reports

Records of firefighter medical evaluations for each incident where rehab is established should be filed and maintained. This is a group record of what findings were made on a certain incident. Should individual records need to be checked to complete a medical history of a member of the fire department, each incident's records could be checked to obtain specific information.

Statistics

Every fire department keeps records and compiles statistics about its entire operation. Often, that collection of data is just a collection of numbers. If you make the effort to collect information, you should make it work for you. Evaluate the records and look at what specifics are in the reports. The following are several examples of records to develop and maintain.

- Annual injury/illness report (department record, IAFF, OSHA Log 200)
- Lost-time report
- Workers' compensation expenses
- Medical exam/therapy expenses

SUMMARY

It can be frightening to think that, even with all of the inherent dangers involved with firefighting and emergency medical work, the leading cause of death on the fireground remains heart attacks. This has been the case for over a decade. Injury rates for firefighters are very high, and it can be inferred that injury rates for EMS providers are also high. Although much has been done in the past 15 years to increase the safety of emergency responders, a focus on health maintenance could perhaps have the greatest positive impact in this area.

Emergency responders have several high risk factors that makes a health maintenance program extremely important. Some of the risk factors that are commonly attributed to emergency responders include the following:

Lack of physical fitness
History of illness
Medication use
Smoking
Hearing loss
Back injuries
Cancer

To combat these risk factors, the International Association of Fire Fighters and the International Association of Fire Chiefs developed the "Joint Wellness/Fitness Initiative."

Other steps that can be taken by a fire department to reduce the risk factors include conducting annual medical evaluations and periodic medical examinations, practicing proper infection control, sanctioning the use of an employee assistance program, and establishing a strong emergency incident rehabilitation program.

Recordkeeping and documentation are important aspects that need to be addressed. Employee medical records files should have documentation of medical evaluations and examinations, rehab reports, vaccination records, exposure and injury reports, and return to duty reports. These records are confidential and must be maintained for the duration of employment plus 30 years.

Good record keeping can lead to good statistics available for statistical analysis of your department's operations. Statistics related to health maintenance can be gathered from injury reports, exposure reports, and medical examination documents, as well as other documents. The summary developed from these reports should be made available to everyone in the department.

7

Accident and Injury Investigation

The investigation of accidents, exposures, deaths, and injuries provides the emergency response agency with information about what went wrong in an accident and how to avoid that situation in the future. Investigations give us facts about an incident that can provide a basis for correcting our SOPs, developing other control measures to prevent the situation from recurring, or minimizing the negative effects of a recurrence.

NFPA 1500 requires the investigation of all accidents, fire apparatus collisions, injuries, fatalities, illnesses, and exposures involving members of the fire department. The standard also mandates that corrective action be taken as a result of the investigation to prevent a recurrence of the situation that led to the loss.

An investigation by a response agency may also play a part in the defense of the agency against any civil or criminal claim made by a member of the public who feels that he or she was injured by an action or lack of action by the response agency. Although others are sure to investigate the incident, the early gathering of facts and photographs of the incident scene will assist the response agency in determining the course of action to take.

Accident investigation may also assist the HSO in discovering trends that lead to accidents. A seemingly unrelated series of accidents may have one root cause. The HSO should be able to determine the cause of most accidents and to recommend changes or enhancements to prevent the situation from recurring.

What Should Be Investigated?

All accidents, vehicle collisions, exposures, illnesses, injuries, and fatalities must be investigated. It also is useful to look into the causes of "near misses." These are accidents that almost happen, such as a brush with death at an intersection during the response. Microseconds can separate a near miss from a tragic accident. Although the reporting associated with accidents usually is not required for near misses, the HSO will be better off if these incidents are reviewed. There is no need to wait for an accident to make a change for the better.

In many agencies, the HSO will be the primary investigator of all accidents. In larger agencies where the response of a single individual is not reasonable, battalion chiefs or field supervisors should be trained to perform routine investigations.

The HSO still should review all accident reports. For significant accidents or exposures, or in the event of the death or serious injury of a responder, the HSO should be involved directly.

The HSO may choose to be involved in any accident investigation, even those that are considered routine. In cases where the HSO or ISO needs help with an investigation, the fire department's arson investigators may be a valuable resource, because they have training in investigative techniques. An excellent guide to investigating is *Emergency Scene Accident Investigation—A Guidebook for Emergency Service Organizations,* developed by VFIS, Inc.

Vehicle Collisions

One of the most significant and spectacular accidents that can be experienced by emergency response agencies is the collision of an emergency response vehicle with another vehicle or an object. These are the most significant risks that we face, aside from those at the emergency scene.

According to USFA, in 1998 the University of Michigan Transportation Research Institute performed a study of fire apparatus traffic collisions for 1994–1996 for the Freightliner Corporation. The study found that:

> There are 2,472 fire apparatus accidents each year;
> Six occupants of fire apparatus are killed each year;
> 413 occupants of fire apparatus are injured each year;
> 20 percent of fire apparatus collisions result in rollovers;
> 47 percent of fire apparatus collisions are at intersections;
> 73 percent of the firefighter fatalities were from rollovers; and
> 76 percent of the firefighters killed were not wearing seat belts.
>
> *Firefighter Fatalities in the United States, 1998,* USFA, p. 32

Upon arrival on the scene of a crash involving an emergency vehicle, the HSO should ensure that all responders involved in the collision and all civilians who may have been injured have received proper medical care. The HSO should make sure that family notifications have been made and that a member of the response agency is available to transport family members to the hospital if that service is needed or desired.

The HSO may choose to have minor vehicle collisions documented by a second-level supervisor such as a battalion chief. In smaller agencies, where there are no on-duty supervisors, the HSO may have to make arrangements to be available or to have some other member of the agency available for response. In any case, the HSO should review all collision reports, minor and major.

Vehicle collision analysis is an art and a science best practiced by those who have occasion to use the skills on a regular basis. The HSO should have a basic knowledge of vehicle collision analysis, but the actual analysis usually is performed by a law enforcement official. The HSO should be involved in the analysis in a supporting and observing role. It may be wise to have a law enforcement official from outside your government structure perform that analysis: There may be a conflict of interest if the city police department investigates a collision involving a city fire truck or ambulance. A good working relationship with law enforcement officials will make this process easier.

In agencies where a significant number of vehicle collisions occur on a regular basis, the HSO should become more practiced and educated in the skills of accident investigation.

At every collision involving an emergency response vehicle, photographs of the scene and the vehicle should be taken. An example of a post-crash photo is shown as Figure 7-1. Use a digital or 35mm camera, so that good quality reproductions and enlargements are possible. A standard report with a drawing of the scene and statements by all responders and witnesses also should be prepared.

Prior to being placed back in service, all equipment and vehicles should be inspected by a qualified person. If there is any question that some malfunction of

FIGURE 7-1 The HSO will typically be involved in the investigation and/or analysis of department vehicle collisions. This fire department tanker (tender) flipped en route to a fire incident; however, because of the proper use of seat belts, there were no injuries. Photo by Gordon M. Sachs.

equipment or apparatus was a factor in the collision, the equipment or apparatus should be impounded. It may be wise to allow law enforcement officials to impound the equipment or vehicle to avoid the allegation of impropriety.

Responder Injury

Like the situation with vehicle collisions, the first concern that the HSO should have upon arrival on the scene should be to ensure that all responders have received proper medical care (see Figure 7-2). The HSO should make sure that family notifications have been made and that a member of the response agency is available to transport family members to the hospital if that service is needed or desired.

In cases of minor injury, the HSO may have the injured responder and their immediate supervisor fill out accident and injury forms. In many states, time limits must be observed between the occurrence of the injury and the report of the injury to the state department of labor or workers' protection organization.

Prior to being placed back in service, all PPE, equipment, and vehicles that may have been involved in the accident should be inspected by a qualified technician.

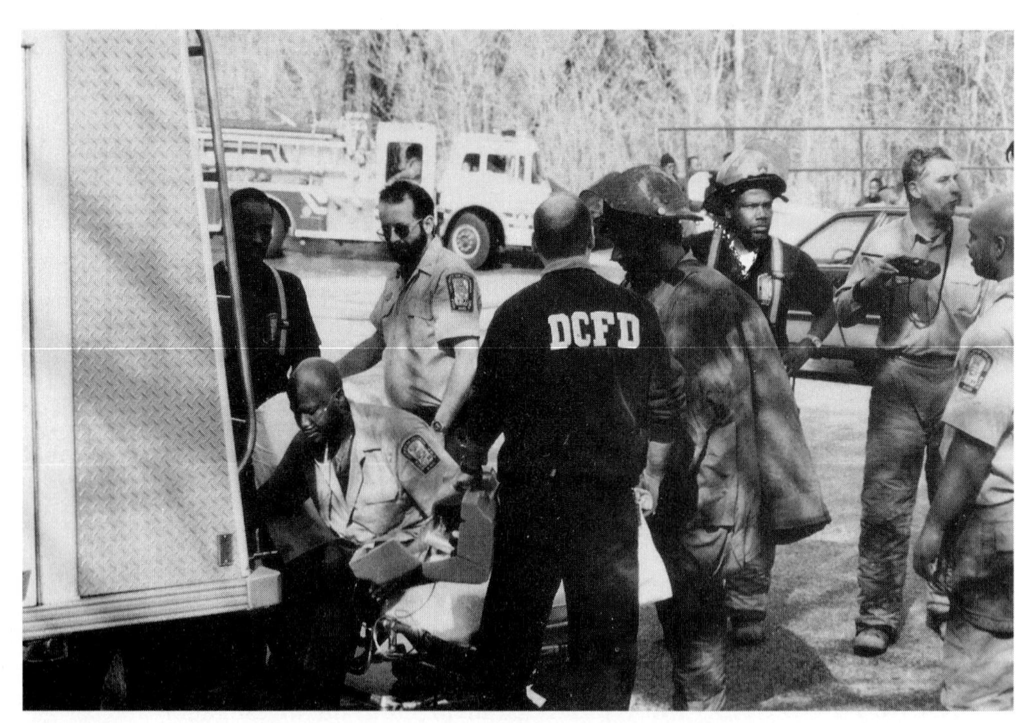

FIGURE 7-2 The HSO will typically be involved in the investigation and/or analysis of on-the-job injuries to responders, whether or not the injuries are incident related. If there is an ISO on scene, the ISO will likely help with any investigation of incident-related injuries. Photo by Louis Carter, Jr., District of Columbia Fire and EMS Department.

In cases of moderate to severe injury to a responder, all PPE and involved equipment should be impounded by the HSO until the conclusion of the investigation.

Responder Fatality

The death of a responder is one of the most stressful events an emergency response agency can experience. The Public Safety Officer's Benefit Program (PSOB) has specific requirements that must be documented prior to any release of benefits. A fact sheet describing the requirements and the benefits of the PSOB program can be found in appendix H. An excellent guide to the investigation of a firefighter fatality is published by the International Association of Fire Chiefs (IAFC). It is recommended that a copy be a part of every fire department's library.

Infectious Disease Exposure

In most situations, these exposures will be documented by the responder who was directly involved. The importance of documenting exposures cannot be understated. A thorough investigation to determine whether the exposure constitutes a significant risk of transmission of disease must be undertaken immediately. Typically, the department's designated infection control officer (often the HSO) will coordinate this.

Managing the Investigation

Accidents and injuries occur routinely under nonemergency operations or conditions. The HSO must manage and control these incidents. This may involve department personnel and civilian members of the department. Policy should dictate notification of the HSO. A response will depend on the nature and severity of the incident.

The company officer, the employee's supervisor, or the ISO may investigate accidents and injuries that occur at emergency scenes. The HSO's and/or the ISO's involvement will depend on the nature and severity of the incident.

A standard organizational form or department reporting procedure must be used when the HSO or ISO conducts an investigation. This report ensures that no information is overlooked or forgotten during the process. A completed written report is forwarded to the appropriate personnel inside and outside the department.

The investigation report may include recommendations for revisions to, or development of, procedures pertaining to department operations. It may indicate that procedures should be developed based on actions at an incident. Based on observations of the ISO at an incident, or due to an accident or injury at an incident, a recommendation for policy development or revision of current policy may be part of the report summary.

The recommendations may indicate deficiencies or inadequacies in department operations. If they relate to training and education needs, the HSO must confront these issues.

Health and Safety Officer's Responsibility

As the program manager of the safety and health program, the HSO has specific duties and responsibilities as part of the post-incident evaluation and analysis. This is an important function, because negative aspects of an incident or situation tend to be overlooked or forgotten quickly. The HSO can make a positive impact under these circumstances.

Interfacing with the Incident Safety Officer

The HSO must assist the ISO as needed during investigation or analysis of an incident. The nature or severity of the incident will dictate the involvement of the HSO. If a firefighter fatality occurs, the HSO will be in charge of the investigation. The ISO may conduct the initial portion of an investigation, and the HSO will take control of the investigation at a designated time.

The HSO may also assist with the investigation if an additional resource is needed for any reason. The HSO may assist to ensure that documentation is complete. If additional information needs to be obtained from department personnel or from a witness, the HSO would assist with this process.

The HSO may be responsible for having the investigation report finalized in a standard report form. If the report needs information from other sources, the HSO may be the focal point of this process.

Record Keeping

The HSO is responsible for information relating to safety and health. This includes reviewing accident and injury reports and providing a monthly or annual analysis. Trends or patterns relating to accidents, injuries, or recurrence of safety and health issues need to be tracked and identified by the HSO.

A copy of every accident and exposure report should become a part of the agency's permanent file. Health records, such as communicable disease exposure records, are required to be retained for 30 years after the responder has left employment with the agency.

The results and recommendations of every investigation should be passed on to the member of the organization who is responsible for that operation. For instance, recommendations concerning changes to the design of fire apparatus should be routed through the chain of command to the agency's apparatus officer.

Discipline, if taken as a result of an accident, should not be the responsibility of the HSO. Although the HSO most likely is an officer in the agency, direct discipline by the HSO may undermine future safety efforts. The HSO should report the facts through the chain of command and let the direct supervisor of the affected responder determine appropriate action.

Risk Management

The HSO is the department's risk manager. Most departments spend between 5 and 15 percent of their time at emergency incidents. The other 85 to 95 percent of the time is spent in nonemergency situations. The HSO is responsible for managing all situations, including vehicle accidents, accidents or injuries at the station or department offices, or other safety and health issues that develop.

During emergency incidents, the HSO may be involved, or the ISO or the incident commander (IC) may act as an on-scene risk manager. The intent is to ensure that the process is covered at each emergency incident.

Investigating

The HSO may or may not be required to respond to investigations. For nonemergency situations, the HSO has the primary responsibility for conducting investigations, based on the nature or severity of the incident. In some cases, the immediate supervisor may be able to conduct the investigation and forward the report to the HSO. However, apparatus or vehicle accidents under nonemergency conditions or personal injuries during daily work functions may require assistance from the HSO.

Emergency investigations may be conducted by the ISO. The nature and severity of the situation will dictate the involvement of the HSO. If a fatality or serious injury occurs, the HSO becomes the primary investigator. In many cases, outside expertise may be needed (see Figure 7-3).

Data and Trend Analysis

The HSO should regularly review data and reports in summary form to determine if any underlying trends emerge. Sometimes, the only way to identify a trend is to look at all data in one place at one time. Looking at reports throughout the month or throughout the year in ones and twos will not allow for proper analysis.

Based on the investigation reports and analysis, the HSO may determine that a training need exists. This could include training and education in both emergency and nonemergency situations. The HSO may conduct the training or ensure that other department staff members conduct it.

After a review of reports and documentation concerning an incident, or having participated in a post-incident analysis, the HSO must evaluate the affected procedures. The HSO may determine that procedures need to be developed or revised to prevent recurrence.

FIGURE 7-3 The HSO may need to return to the scene of an accident or injury during an investigation. The HSO must often turn to technical experts, such as accident reconstruction specialists, structural engineers, or fire investigators for assistance. Photo by Gerry Suftko, Mesa (AZ) Fire Department.

Modifications

The HSO may be required to evaluate the status or selection of personal protective equipment and clothing, apparatus, and facilities.

Issues concerning personal protective clothing may include whether the clothing was being used properly; if it failed, why; discussions with the manufacturer; and whether the proper clothing was being used for the proper incident (e.g., hazardous materials and infection control). Personal protective equipment must be examined thoroughly to determine if a problem exists. This may require involving the manufacturer or a testing laboratory to assist. Does the problem exist due to poor preventive maintenance or improper specifications? The investigation process will help determine what modifications are needed.

The HSO will have to work with department mechanics to determine problems with apparatus. Does the problem exist because of human error or because of poor preventive maintenance? After the problem is identified, the source may be traced to training and education, as well as to procedure development.

Facilities issues may evolve from new laws, standards, and regulations (e.g., OSHA bloodborne pathogens regulations or the Americans with Disabilities Act) or from poor design or construction (e.g., sprinkler system malfunctions). An ongoing risk management process will ensure that facilities are upgraded to meet current regulations.

The Occupational Safety and Health (OSH) Committee's Responsibilities

If the department does not have an HSO, the occupational safety and health (OSH) committee may be responsible for managing the safety and health program. The responsibilities may be distributed so that each member or small group of the OSH committee is assigned a particular function, based on expertise or interest.

With an HSO directing the department's safety and health program, the OSH committee still plays a vital role and is a valuable resource to the HSO. The OSH committee can assist with training, development, or revision of procedures; rewrite modifications; or help with any other situation to complete a project and improve safety and health.

SUMMARY

The investigation of accidents is one of the most important duties of the HSO. Problems identified during this process will enhance the safety of all responders.

If an accident or near miss occurs and no action is taken to correct the conditions that contributed to the incident, it will happen again. The only unknown factor is when it will happen.

Part Three

Operational Aspects—
The Incident Safety Officer

8

Emergency Incident Risk Management

As discussed in chapter 5, risk is a part of every day for each of us. Every action that we take carries with it the chance that we might be killed or injured. Risk has an impact on us as we work and as we play; there is no way to completely escape it. No activity is completely without risk.

In most cases, the risks that we take are minor. The risk of being hit by lightning is very low, yet few of us are comfortable in an open field during an electrical storm. Likewise, the risk of crashing in a commercial airliner is minor but always on everyone's mind during takeoff and landing. We each face risks in our lives at work and away from work every day. Sometimes, we do not think about the risks that we face, but they are still there.

Risk management is the method used to reduce exposure to risks. Many risks in the emergency services cannot be completely avoided; fire fighting, emergency medical care, and special operations are extremely hazardous activities. Therefore, risk management for emergency response organizations is divided into two categories: pre-emergency risk management and emergency incident risk management.

Interfacing Pre-Emergency and Emergency Incident Risk Management

Pre-emergency risk management, used prior to emergency situations, is based on the classic risk management process discussed in chapter 5. These activities can have a major impact on the safety of members working at the scene of an emergency. Pre-emergency risk management is the responsibility primarily of the fire department health and safety officer (HSO), although the incident safety officer (ISO) must know the techniques used and the products of this type of risk management.

Some of the most dangerous nonemergency activities that responders can perform are live-fire training, multicasualty exercises, extrication training, and training for special operations, such as hazardous materials emergencies (see Figure 8-1) or high-angle rescue. Risks at these training exercises mirror the risks in emergency operations. Although the HSO may be responsible for developing a training safety plan, an ISO should be operating during each evaluation. The ISO should apply the same risk management techniques used at emergencies to these events.

FIGURE 8-1 Many training evolutions have the same amount of risk as "the real thing" and should be treated in the same way. Live fire exercises, extrication training, and special operations training are some examples when a safety plan should be developed, an ISO in place, and risk management practices used throughout the evolution. Photo courtesy Chesterfield County (VA) Fire and EMS.

NFPA 1403, *Standard on Live Fire Training Evolutions in Structures,* provides safety and operational guidance for live structural fire training and should be used as a guide for the HSO and the ISO.

Emergency Incident Risk Management

In contrast to the studied approach of pre-emergency risk management, risk management at emergencies is a constantly changing, fast-paced activity. Although the HSO is the risk manager for all pre-emergency risk management measures, emergency risk management is the primary job of the ISO. Safety cues will assist the ISO.

The ISO's Role in Emergency Risk Management

The ISO's first task as the risk manager for emergency scenes is to ensure that all pre-emergency risk management measures are being followed by responders. PPE and infection control equipment are of no value unless they are used at the emergency.

The ISO should actively survey the scene of the emergency and make sure that all members engaged in operations are properly protected. Other pre-emergency risk management measures—such as the use of cold, warm, and hot zones for hazardous materials emergencies—must also be in place.

The ISO must watch constantly to make sure that safety equipment is in place and that safety procedures are being followed. Emergencies are dynamic events, so the ISO must continually monitor the emergency scene. A scene that is safe at one moment may not be safe half an hour later.

The ISO and the Incident Commander (IC)

The IC of an emergency has overall responsibility for the safety of responders working on the scene of the emergency. That responsibility cannot be delegated. The ISO assists the IC and acts as the IC's eyes and ears on matters related to responders' safety.

The ISO reports directly to the IC as a part of an incident management system. This direct access to the IC allows the ISO to transmit information directly to the IC without the fear that the message will be scrambled in translation or overruled by a tactical-level officer. This position reflects the importance of the ISO's role at the emergency scene.

Every responder at the scene of the emergency has a safety responsibility. Each individual member has a duty to be safe, to watch out for the safety of other responders at the scene, and to cooperate with safety procedures. The IC depends upon the ISO to monitor the safety of the scene; this responsibility may require the ISO to issue corrective instructions to responders at the emergency scene. This does not mean that the ISO needs to assume the role of safety cop (see Figure 8-2). Safety on the emergency scene is too important to be turned into a game of cop and robber. If problems are noted, the ISO should correct them in the simplest way possible. The safety of responders should be the key, not an attempt to catch and punish errors.

FIGURE 8-2 The Safety Officer Is Not a "Safety Cop."

The ISO's Duties at the Emergency Scene

Each emergency response agency should have a written policy that outlines the response criteria for an ISO. An ISO is not needed at every incident; using an ISO should be reserved for more significant incidents.

The use of an ISO is recommended at working structural fires in commercial buildings; multicasualty incidents; all multiple-alarm fires; and all special operations incidents, such as trench rescues, water rescues, hazardous materials incidents, and high-angle rescues. The use of an ISO is required by federal law at all hazardous materials incidents. The use of an ISO is also strongly recommended for incidents that involve multiple responder injuries or the death of a responder.

The ISO's only job at the scene of an emergency is to watch over and monitor the safety of responders. The ISO cannot be tied down to a single location, because the nature of emergencies is always changing. The assignment of other duties or sectors to the ISO is not appropriate, given the risks faced by responders at the emergency scene. The ISO must look for risks of immediate danger to responders and continuously evaluate the emergency scene for risks that may present danger in the near future.

The ISO's Knowledge of Risks

The ISO is a member of the emergency response organization who is well versed in the procedures of the agency and the dangers that are present at emergency scenes. Most agencies will not designate an ISO prior to an emergency. Very few agencies have the staffing and emergency call volume necessary to make an on-duty ISO a reality. Some volunteer organizations may designate an individual to serve as an ISO, but other members of the department may fill the position if the designated ISO is unavailable.

The agency's HSO may well perform the duties of ISO, but the ISO may also be chosen by the IC at the scene of an emergency from the members of the agency who are available to serve. The IC must take care to select an individual who possesses the experience, knowledge, and ability to perform the functions of an ISO adequately.

The IC and all of the responders at the scene of the emergency count on the ISO, whether designated beforehand or chosen at the scene of emergency, to keep a safety focus. The ISO relies on prior experience working at emergencies, training in emergency operations and safety, safety cues, and intuition to watch for risks to responder safety and predict or forecast what will happen at the emergency if operations continue.

Any type of incident—fire, EMS, Hazmat, or technical rescue—may dictate the need for an ISO to respond or be appointed. A few individuals may need to be trained to act as the safety officer; this will ensure that someone is available at all times. An effective ISO understands that the primary responsibility of an ISO is *not* that of an enforcer who dictates to others. Rather, the ISO's duties at any incident (EMS or fire) are to support the incident commander and the incident management system to ensure that safety is an important component of the incident action plan and that the responders are operating as safely as conditions at the incident allow. This person, however, must have the authority to stop unsafe acts.

An effective ISO for EMS incidents must know the dangers that may be present during confined-space rescues, hazardous materials emergencies, and mass-casualty incidents. Some examples of knowledge required include infection control procedures; scene security measures; personal protective equipment requirements; critical incident stress indicators; the types of safety lines needed at water rescues; shoring needed at trench collapses, and as shown in Figure 8-3, safety concerns at structural collapses.

The individual designated as ISO at these and other incidents must understand personal limitations and recognize that no one individual can be expected to be an expert in every facet of every type of emergency incident. Not only should an ISO get help when needed from a technical specialist or outside expert, but the ISO also can have the incident commander appoint others to assist with the safety function at large incidents, when necessary.

FIGURE 8-3 An effective ISO must know the dangers that may be present at various types of incidents; however, the ISO does not need to be a technical expert in each area. The ISO must be able to identify and act on safety cues, and must know when to get help—either assistant ISOs or technical experts. Photo by Steve Weissman, Fairfax County (VA) Fire and Rescue Department.

Forecasting

Given the fact that emergency scenes are dangerous places where situations change rapidly, the ISO must monitor the safety of responders constantly. The ISO must be concerned with hazards or risks that present an immediate danger to responders or that may become dangerous to responders. The ISO must *forecast* the future of the emergency as it relates to responder safety.

No member of the emergency services, not even the ISO, has a crystal ball that can be used to predict the future with 100-percent accuracy. The ISO must use experience, training, safety cues, and intuition to get ahead of the emergency and predict developments that will have an impact on the safety of responders. An example of this is shown in Figures 8-4a and 8-4b—an experienced ISO would recognize the cues indicating that the second-floor room in the front is about to "flash."

FIGURE 8-4a It is important for the ISO to be able to identify safety cues. The photo above has many—one of which is that the second floor room in the front is about to flash (as seen in Figure 8-4b). Photo courtesy Gordon M. Sachs.

FIGURE 8-4b Photo courtesy Gordon M. Sachs.

Forecasting Tools

A television meteorologist uses information provided by the National Weather Service, local radar, and satellite photographs to predict or forecast weather. Likewise, the ISO has tools that can be used to predict the future of an emergency. The following tools are intended to assist the ISO with structural fires, emergency medical incidents, and special operations incidents.

Structural Fire Forecasting Tools

At a structural fire, the ISO's forecast must take the following into account.

- The features of the fire building.
  - Access to the interior may be difficult for firefighters, as seen in Figure 8-5.
  - Maze-like floor plans increase risks.
  - Utilities that are provided to the building, such as gas, electricity, or steam.

FIGURE 8-5 The ISO must be able to look at the "big picture" and recognize the features of the building and the effects of the fire that might jeopardize the safety of firefighters operating inside. Photo by Louis Carter, Jr., District of Columbia Fire and EMS Department.

- Fire protection systems.
  - Operating sprinklers indicate a working fire.
  - Added weight from sprinkler water eventually may cause structural problems.
  - Automatic or manual smoke vents may assist firefighters working in interior positions.
  - The presence of special agent systems, such as dry chemicals and Halon, indicates a special hazard is present. These systems usually are designed to discharge only once.
- Access for fire crews.
  - Working in large buildings causes fatigue problems.
  - If the fire appears to be in hidden spaces, opening up for suppression will be time-consuming and will cause firefighters to become fatigued.
- Egress for crews working on the interior.
  - Crews must be able to find their way out if an emergency occurs.
  - Ladders to upper-story windows provide alternative escape routes.
  - Stairways must be intact, or alternate means of egress provided as shown in Figure 8-6.
- Construction type.
  - Bowstring truss and lightweight truss roofs often fail early, as occured in Figure 8-7. Their failures can have dramatic and deadly consequences.

FIGURE 8-6 Access to and egress from a building or any part of a building are critical firefighter safety factors. Sometimes access/egress problems can be overcome while maintaining an acceptable level of safety. Photo Steve Weissman, Fairfax County (VA) Fire and Rescue Department.

FIGURE 8-7 Lightweight wood-frame construction can mean early collapse or rapid fire spread—both risky for firefighters. Photo courtesy Chesterfield County (VA) Fire and EMS.

  - Look for the presence of "stars" and other indications that the structure has been reinforced by steel rods.
  - Look and listen for early signs of structural failure, e.g., groaning, smoking mortar, or bulges.
  - Look for other construction hazards, such as suspended loads.
- Age of the fire building.
  - Older buildings generally do not have lightweight trusses.
  - Egress and access may be difficult.
  - Look for signs of structural weakness, such as reinforcing rods.
  - A new building is just as likely to collapse as an older building.
- The potential for fire extension into exposed buildings.
  - Gauge the amount of fire involvement.
  - Distance between buildings.
  - Wind conditions.
- Amount of fire involvement.
  - Big fires usually mean no survivors and less benefit from risk to firefighters engaged in interior operations.
  - Lots of fire involvement leads to early structural failure, as may be indicated by cues in Figure 8-8.

FIGURE 8-8 Heavy fire involvement often means no survivors and greater risks to interior firefighters. Older buildings may have heavier construction techniques, but many also have inherent structural weaknesses. Risk versus benefit is an important consideration in this type of situation. Photo courtesy Gordon M. Sachs.

- Roof hazards.
  - Ladders at two corners allow for escape.
  - Firefighters' walking on structural members is a hazard.
  - No "roof shepherds"; once the hole is cut, get off the roof.
  - If the fire is well vented, no hole is needed.
  - Watch for potential collapse, bowstring trusses, or lightweight wood trusses.
- Time.
  - Time from ignition to flashover may be as little as 2 or 3 minutes. The fire in Figure 8-9 had only been burning a few minutes prior to the firefighters' initial attack.
  - The longer the fire burns, the weaker the structure becomes.
  - Taking into account the time interval between the arrival of the first unit on the scene and the response time of the ISO, it is possible that the incident has been going on for much longer than the ISO has been there.
  - Time can work against emergency responders.

FIGURE 8-9 One unknown common to fire incidents is the length of time the fire has been burning. The longer the fire burns, the weaker the structure becomes. What looks like a "simple" room-and-contents fire may be the precursor to a collapse. Photo courtesy Chesterfield County (VA) Fire and EMS.

- The IC's tactical objectives.
  - If things do not go as hoped, firefighters must be able to remove themselves from the hazard area.
  - Water supply (hydrants or tanker operation) must be adequate to support safe operations. Insufficient water, or running out of water, can put firefighters in a dangerous position. Look for hydrants or recommend a tanker operation.
- The weather.
  - Extreme heat dictates the early initiation of rehab and more frequent work/rest cycles.
  - Cold weather presents hazards in addition to hypothermia, e.g., slippery surfaces and mud (see Figure 8-10).
  - Electrical storms may create lightning hazards.
  - Ground ladders may be blown over by strong winds.

Medical Emergency Forecasting

At a medical emergency, the ISO's forecast must take the following into account.

FIGURE 8-10 The weather—hot or cold—can affect operations by limiting the capabilities of firefighters. Rapid and more frequent rehab may be required. Cold weather may add the risks of slippery surfaces and obstacles hidden in the snow. Photo by Robert Rosensteel, Sr., Vigilant Hose Company, Emmitsburg, MD.

- Protection from communicable diseases.
- Protection from physical hazards, such as sharp surfaces. All members operating in or around a vehicle during extrication should use full PPE, as shown in Figure 8-11.
- Violent acts.
  - If a crowd has gathered, individuals may become agitated or violent toward responders.
  - The person who committed the violent act may still be in the area.
  - Escape routes will be needed if the situation worsens.
  - Law enforcement presence helps to ensure responders' safety.
- Protection from surroundings.
  - Moving traffic—emergency vehicles should be used to shield from traffic, if possible, as was done during the incident shown in Figure 8-12.
  - Car/pedestrian accident—move patient onto curb and out of traffic for treatment, if possible.
  - Weather—heat, cold, rain, or sleet.
- Sufficient staffing.
  - Staffing must be adequate to carry and load the patient(s) into the ambulance.

FIGURE 8-11 Vehicle crashes—especially those involving extrication—inherently have sharp surfaces. Often, EMS personnel are not wearing personal protective equipment to protect themselves from these hazards. The ISO should move the unprotected EMS personnel away from the vehicle, or provide them with protective equipment. Photo by Orlando Dominguez, Brevard County (FL) Fire/Rescue.

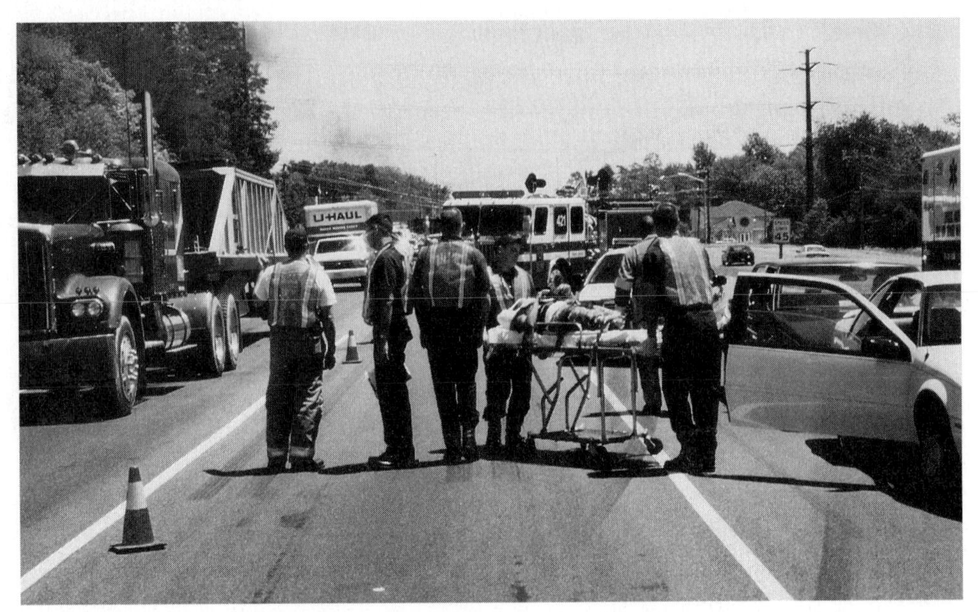

FIGURE 8-12 Apparatus should be used as a shield whenever personnel are working on a roadway. Reflective traffic cones and reflective garments should be used to increase visibility at both daytime and nighttime incidents. Photo by Steve Weissman, Fairfax County (VA) Fire and Rescue Department.

- Help may be available from bystanders or law enforcement officials.
- If the patient is far from the road or from a paved surface, more people will be needed to carry the gurney.
- More patients indicate the need for more responders, as shown in Figure 8-13.

Special Operations Forecasting

At special operations incidents, the ISO's forecast must take the following into account.

- Incident duration will be longer.
- Technical experts.
  - Technical experts should be present at the scene, as shown in Figure 8-14.
  - Technical experts must be capable of ensuring the safety of responders engaged in special operations.
- Properly equipped responders, such as the one shown in Figure 8-15, are available for rescue of members in hazardous areas.
- Time generally is less of a factor, because more time is available to prepare for action.
- Time also can be an enemy to responders who may drop their guard and be less aware of hazards as the incident drags on. Figure 8-16 shows a dump fire—one type of incident where this is a major factor.

FIGURE 8-13 There should be sufficient staffing on hand to safely move each patient from the scene to the treatment area or ambulance. For incidents where a patient (or patients) must be carried a long distance, several crews may be needed to trade off the carrying tasks. Photo by Steve Weissman, Fairfax County (VA) Fire and Rescue Department.

FIGURE 8-14 The ISO should not hesitate to consult technical experts from within or outside the department. This is especially true at hazardous materials incidents (e.g., chemists) or technical rescue incidents (e.g., structural engineers). Photo by Ed Roberts, Dothan (AL) Fire Department.

FIGURE 8-15 Time is less critical at haz mat and technical rescue incidents. All necessary precautions, including the use of the proper level of PPE and the establishment of an appropriate decontamination area, must be taken. Photo by Gerry Suftko, Mesa (AZ) Fire Department.

Rapid Intervention Crews (RICs)

In many situations, such as structural fires and hazardous materials incidents, responders operating in the hazardous area may get into situations where they need to be rescued. This might be the case at any structural fire or hazardous materials incident where crews cannot quickly extricate themselves from the hazard area if something goes wrong.

FIGURE 8-16 Fires involving dumps or landfills will be long exhausting events with many obvious and latent hazards. Photo by Gordon M. Sachs.

For fire and special operations emergencies, OSHA regulations and NFPA 1500 require that the IC evaluate the hazards and risks of every emergency situation and assign RICs as needed. NFPA 1500, Section 6-5, is devoted to the topic of "Rapid Intervention for Rescue of Members." RICs are crews of at least two properly equipped responders who are available to perform a rescue of other responders, if required. The ISO should use forecasting skills to assist the IC with information on how many and what type of RICs are needed for an emergency.

Initially, RICs may be made up of responders dedicated to that mission; or, early in the incident, they may be made up of responders on the scene of the emergency who are performing other tasks but are ready to redeploy and perform RIC functions (see Figure 8.17). As the incident progresses in size or complexity, RICs will either be made up of responders on the scene dedicated to the task or be a company or companies of responders in a staging area or on the scene of the emergency. In cases where responders are in the hazard area of a special operations incident or where responders are in an area where an equipment malfunction could lead to their incapacitation, the activation of at least one RIC is required.

The RICs must be ready to move into the hazard area at any time to perform a rescue. They must be equipped with protective clothing and equipment that will allow them to operate safely. For structural fires, this would include full structural protective clothing, SCBA, and any specialized tools that may be necessary to

FIGURE 8-17 A rapid intervention crew (RIC) is made up of at least two properly equipped responders who are at the ready to assist other responders who may be in trouble. At a working fire, the RIC initially may be performing other tasks while remaining ready to perform RIC functions. As soon as possible, however, a dedicated RIC should be established. Photo courtesy Gordon M. Sachs.

gain access to trapped firefighters. In the case of a hazardous materials emergency, RICs should be equipped with the proper level of protective garment and any equipment that may be needed. RICs should be staged near the command post (CP) so that face-to-face communication with the IC is possible, as seen in the background of Figure 8-18.

Experience on emergency scenes has proved the need for dedicated RICs. The individual most likely to have good information on the location of responders in need of rescue is the IC. When members at the scene of an emergency learn that

FIGURE 8-18 The ISO, accountability officer, and RIC team are important components of the incident commander's safety plan. The dedicated RIC may be staged near the command post for face-to-face communications with the ISO and incident commander, as well as for rapid deployment if necessary. Some incidents may require more than one RIC. Photo by Gordon M. Sachs.

other responders are in need of immediate assistance, their automatic reaction is to rush to the area where they think their help will be needed. Often, the information in their possession is inaccurate, and their efforts are wasted.

If the RIC is located close to the CP, the IC can access the RIC, give them good information on where their help is needed, and send them on their way. Many fire departments use the RIC concept but do not use the RIC name. For example, some departments use "safety companies" or "FAST companies" (firefighter assistance and search teams) to stand by in case the rescue of firefighters is needed. An extra engine is added to the response to perform this function. There is no magic in the name; the concept of responders dedicated to the mission of rescue is more important.

Emergency Medical Support

As a part of the forecasting process, the ISO should attempt to predict the need for EMS support on the scene of both EMS and non-EMS emergencies. In many communities, the local EMS provider routinely responds to all structural fires. If EMS support is not present on the scene, the ISO should forecast the need for this support and request it from the IC, if necessary.

FIGURE 8-19 All incidents should have some level of EMS support, primarily for the benefit of responders. Major incidents, especially fire incidents, may require a large EMS response to support the operation. This EMS response should be a part of the incident command system organization at the incident. Photo by Louis Carter, Jr., District of Columbia Fire and EMS Department.

In some cases, the presence of EMS support is required by OSHA and EPA regulations or by NFPA 1500. In cases where responders are performing special operations duties such as hazardous materials control, the highest available level of EMS must be provided. This support must be at least basic life support (BLS) with medical transportation available. At all other emergency incidents, the IC must evaluate the risks to responders and, if necessary, request that ALS- or BLS-level EMS be standing by at the scene. The ISO should assist the IC with this decision, and should brief the EMS group supervisor of potential hazards or the type of patients that may be encountered (see Figure 8-19).

Once EMS support has arrived on the scene, the ISO should ensure that they are prepared to provide service to responders involved in the incident. If they are pulled into duty for patient care, additional EMS resources should be called to the scene. The EMS crews should be dedicated to establishing and staffing rehab and treatment areas, and they should watch for responders that are too fatigued to continue working (see Figure 8-20). The ISO should attempt to secure the help of higher level EMS providers such as paramedics, because personnel with that level of training can provide the highest level of care for responders who may be injured.

FIGURE 8-20 If the EMS providers are called upon to treat a firefighter, consideration should be given to calling for additional EMS support. This way, if the EMS providers must transport the victim, another EMS unit is in place (or at least on the way) to maintain the on-scene EMS support. Photo by Steve Weissman, Fairfax County (VA) Fire and Rescue Department.

Acceptable Risks

Risks for Responders

Emergency response personnel are action oriented. The excitement and challenges present in the emergency services draw people who are willing to take risks. Fire departments, EMS providers, and other emergency service providers routinely accept risks that would not be acceptable in the private sector.

The acceptance of higher levels of risk does not mean that responders should lay their lives on the line in every situation. The provision of training, protective clothing and equipment, SOPs, the use of ICS, and the use of ISOs minimize risks to responders (see Figure 8-21).

Emergency Incident Risk Management Plan

A risk management plan is a tool for determining which risks are acceptable. Each action on the emergency scene carries with it a benefit and a risk. Acceptable risks

FIGURE 8-21 An acceptable level of risk for responders is based on training, equipment, SOPs, and the use of an incident command system. For example, only responders trained and equipped for swift water rescue, operating under appropriate SOPs and within a standard ICS, should attempt a rescue like this one. Photo by Louis Carter, Jr., District of Columbia Fire and EMS Department.

are those where the benefit is of more importance or value than the negative possibilities posed by the risk. A risk management tool that can be used to help the ISO evaluate risks that are not specifically addressed in the agency's risk management plan is outlined in Figure 8-22. It may sound simple, but the decisions that must be made with this tool are not simple, by any means.

In other words, firefighters and EMS providers should put their own lives at risk in a calculated manner *only* to save a viable patient. They should not take risks

Emergency Incident Risk Management Guidelines

- Emergency responders should risk a lot only to save a lot.
- Emergency responders should risk only a little to save a little.
- Emergency responders should risk nothing to save what is already lost.

FIGURE 8-22 Emergency Incident Risk Management Guidelines.

FIGURE 8-23 Responders should risk their lives in a calculated manner *only* to save a viable patient. While these "risky" operations are underway, other responders should be working to make the situation safer for those and other responders. Photo by Louis Carter, Jr., District of Columbia Fire and EMS Department.

to attempt to save someone who is obviously deceased or property that has already been destroyed. Consider the scene in Figure 8-23. Think about the outcomes, not just the actions!

Unacceptable Risks

Some risks are clearly unacceptable, even for emergency responders. This fact should be addressed in the response agency's risk management plan.

Although the primary responsibility for the development of this plan usually is placed with the HSO, members of the organization who may serve as ISOs should have input in the plan's development. In many cases, the HSO may act as the ISO at particular emergency incidents. At other times, other members of the response agency serve as the ISO. In any case, the ISO on every incident must interpret and apply the agency's risk management plan. Personal judgment on the part of each ISO will play a major part in this ongoing activity.

When a Risk Is Unacceptable

There is no reason for risking the life of an EMS responder in a situation where the injured person is inaccessible because of gunfire or some other extreme hazard.

FIGURE 8-24 Sometimes, the risk is too great, even for the highest trained and best equipped firefighters. Responders should risk nothing to save what is already lost. Photo by Louis Carter, Jr., District of Columbia Fire and EMS Department.

There is no reason for risking the life of a firefighter while mounting an interior attack on a fire in an unoccupied building. Attempting to reach the patient in the vehicle shown in Figure 8-24 prior to the electricity being disconnected would be considered an unacceptable risk.

Not all risk management decisions are this simple. We live in a complicated world, and there is a fine line between an acceptable risk and an unacceptable risk. Bad information or a lack of information about the emergency can blur this line further.

FIGURE 8-25 Sometimes, the incident commander makes a decision without having full knowledge of all aspects of the situation. In this case, this attack crew would not have been allowed to get into this position if it was known that the fire would envelop them. The ISO must be prepared to take the necessary action to prevent this type of situation—sometimes by suspending operations long enough to inform the incident commander of the situation. Photo by G. Emas, Edmonton (Alberta) Emergency Response Department.

Although the IC has overall responsibility for the safety of all responders on the scene of an emergency, they depend on the ISO to help with this critical obligation. The ISO must evaluate the situation continuously to ensure that the IC is aware of the risks of an operation and the consequences if something goes wrong.

In some cases, the IC may make a tactical decision without being aware of all of the risks that responders will face. Figure 8-25 shows such a situation. Some command decisions are made with good information and full awareness of the risks to be faced. In some situations, the IC may make a decision to take a significant risk because of the significant benefit that will occur if the decision pays off.

The ISO's primary concern is the safety of responders working at the scene of an emergency. If the ISO believes that an operation, or any part of it, poses an unacceptable imminent danger to responders, the ISO has the authority to alter, suspend, or terminate an operation or parts of the operation.

Terminating Unsafe Operations

The ISO's authority to alter, suspend, or terminate operations that present an imminent safety hazard to emergency responders is contained in *Standard on* NFPA 1521, *Fire Department Safety Officer,* and paragraph (q) (3) (viii) of the OSHA *Standard on Hazardous Waste Operations and Emergency Response* (HAZWOPER) regulation (29 CFR 1910.120, Final Rule).

NFPA 1521 Section 2.5.1 reads:

> At an emergency incident where activities are judged by the incident safety officer to be unsafe or involve an imminent hazard, the incident safety officer shall have the authority to alter, suspend, or terminate those activities. The incident safety officer shall immediately inform the incident commander of any actions taken to correct imminent hazards at an emergency scene.

The OSHA *Hazardous Waste Operations and Emergency Response* regulation reads:

> (q) (3) (viii) When activities are judged by the safety official to be an IDLH (immediately dangerous to life and health) and/or to involve an imminent danger condition, the safety official shall have the authority to alter, suspend, or terminate those activities. The safety official shall immediately inform the individual in charge of the ICS of any actions needed to be taken to correct these hazards at the emergency scene. (Parenthetical phrases added for explanation.)

Although the ISO does have the authority to alter, suspend, or terminate an activity, this authority must be balanced between the need to protect the lives of responders and the ISO's role in the incident management system. The ISO is a support officer for the IC. The IC counts on the ISO to look out for the safety of responders and to inform the IC of situations that present risks to responders. In most cases, when the ISO spots a risk to the safety of responders, the ISO can take minor corrective action to address the problem (like asking a responder to wear gloves or coordinating with a sector officer to establish a collapse zone).

The situation where an ISO would terminate an operation without first consulting with a division/group/sector officer or the IC is extremely rare and reserved for hazards that place responders in imminent danger. One example of this was a crash and fire involving a propane delivery truck, shown in Figure 8-26. When the ISO arrived on the scene, he observed numerous firefighters operating hand lines and performing other tasks within about 200 feet of the tanker, which was venting from the pressure relief valve. The incident was in a rural area with no life hazard (other than the firefighters), and no structures within one-quarter mile. The ISO made an immediate decision to terminate the operation and evacuate the area of all firefighters. Based upon the emergency incident risk management guidelines, this was a situation where little, if anything, needed to be risked by the firefighters, yet they had been placed in a position of high risk. After conferring with the IC, the ISO agreed that placing two firefighters in a temporary high-risk situation to set up an unmanned monitor was acceptable; this was done

FIGURE 8-26 At an emergency incident where activities are judged by the ISO to involve an imminent hazard, the ISO has the authority to alter, terminate, or suspend those activities. Photo by Gordon M. Sachs.

FIGURE 8-27 The ISO must have a good relationship with the incident commander, operations section chief, and other members of the ICS organization. That way, a decision to alter, suspend, or terminate an operation will be taken seriously and without question. Photo by Ed Roberts, Dothan (AL) Fire Department.

with a RIC team and protective hoseline in place. Once the monitor was charged, the personnel withdrew and the situation was handled safely from that point on.

If the ISO decides that it is necessary to terminate, suspend, or alter a significant part of an operation without the prior approval of a division/group/sector officer or the IC, the ISO should be able to defend that action to the IC (see Figure 8-27). A confrontational relationship between the ISO and the IC should be avoided, if at all possible. The ISO is a support officer to the IC, and the IC may choose to reverse the decision of the ISO and continue an operation. It is in the best interest of the ISO, the IC, and the responders working on the scene to keep communications between the ISO and the IC positive and supportive.

The decision to alter, suspend, or terminate an operation or a part of an operation must not be taken lightly. The ISO must consider the impact of this action on the rest of the emergency operation. Termination of one part of an operation may place responders operating in other areas of the emergency in great danger. The ISO must relay the decision to terminate an operation to the IC as soon as possible.

These decisions tend to be significant events that will be remembered for a long time by both the IC and the ISO. The ISO must make sure that the decision is valid, but they cannot engage in hours of discussion and consideration. If time permits, the safety of the overall operation may be improved if the ISO can consult face-to-face with the IC prior to terminating an operation. The ISO's only job is responder safety; career considerations must take a back seat to the primary function of the ISO—safety.

SUMMARY

The ISO is the on-scene risk manager. The ISO uses the agency risk management plan, risk management techniques, training, experience, safety cues, and intuition to perform their job. Although the ISO has a responsibility for the safety of responders, he or she must operate as a support officer for the IC. The ISO must look for immediate risks and forecast risks that may threaten the safety of responders. The ISO has the authority to alter, suspend, or terminate an unsafe operation. Good communications between the ISO and the IC help assure the safety of responders.

9

Operating at Emergency Incidents

Communications play a vital role in the overall process for the ISO, from the time of notification, through arrival, until the conclusion of the incident. The better the communications, the better the ISO will be able to function.

If the ISO is dispatched automatically to a working incident, the response should be per department protocol. The ISO notifies the dispatcher that he or she is responding to the working incident. The dispatcher may relay information from the IC or have other pertinent information for the ISO.

Arrival/Appointment of an ISO

As the ISO arrives, they must put on appropriate personal protective equipment (PPE). For example, at a working structure fire the ISO will need full protective clothing plus self-contained breathing apparatus (SCBA) and facepiece. The ISO then proceeds to the command post (CP) for a face-to-face meeting with the IC.

The meeting with the IC must be done in person to allow the IC to give direct orders to the ISO (see Figure 9-1). If there are any questions or concerns on the ISO's part, they can be addressed at this time. The IC is busy managing the incident and needs as few distractions as possible. A face-to-face meeting will also reduce the amount of radio traffic and transmissions, as radio channels are usually busy during an incident.

The department accountability policy should require that the ISO check in or drop off a passport or accountability tag at the CP. This ensures that the IC knows the ISO is present.

If the IC appoints a qualified staff member as ISO, the IC will need to brief this individual properly and to relay any strategies or concerns to them to monitor.

If the arriving ISO is given the safety assignment, the acting ISO must brief the arriving ISO at the CP. This will ensure a proper exchange of information. As needed, the IC may be part of the briefing.

Communications Throughout the Incident

The ISO is part of the command staff. The ISO must maintain communications with the IC, usually by listening to assignments given during the course of the

FIGURE 9-1 The ISO must meet face-to-face with the incident commander to ensure that the incident action plan is understood by the ISO and that the safety concerns of the incident commander are addressed appropriately. Photo by Gerry Suftko, Mesa (AZ) Fire Department.

operations, or from personnel communicating with command (see Figure 9-2). A lot can be gained by listening and observing.

If the ISO observes a safety concern, they must notify the IC either immediately or after ensuring the safety of personnel, equipment, and/or apparatus. This may affect the IC's incident action plan, so communication is important. The ISO can advise the IC of problems or concerns involving the following.

- Condition of personnel, especially personnel that are tired and need rehab
- Change in conditions of a structure, such as a partial collapse
- Other problems, such as accountability issues or freelancing

As operations at an incident are being terminated, the ISO and the IC again meet face-to-face. At this time, the ISO provides a briefing on the positive and negative aspects of the incident. The IC will need this information for the post-incident analysis (PIA). The ISO may be required by department policy to submit a written report. Reviewing incident documentation (such as that shown in Figure 9-3) may be beneficial.

If the ISO is removed or reassigned during the incident, a briefing must take place to ensure an adequate exchange of information. Issues that need to be addressed are documented and presented at the PIA.

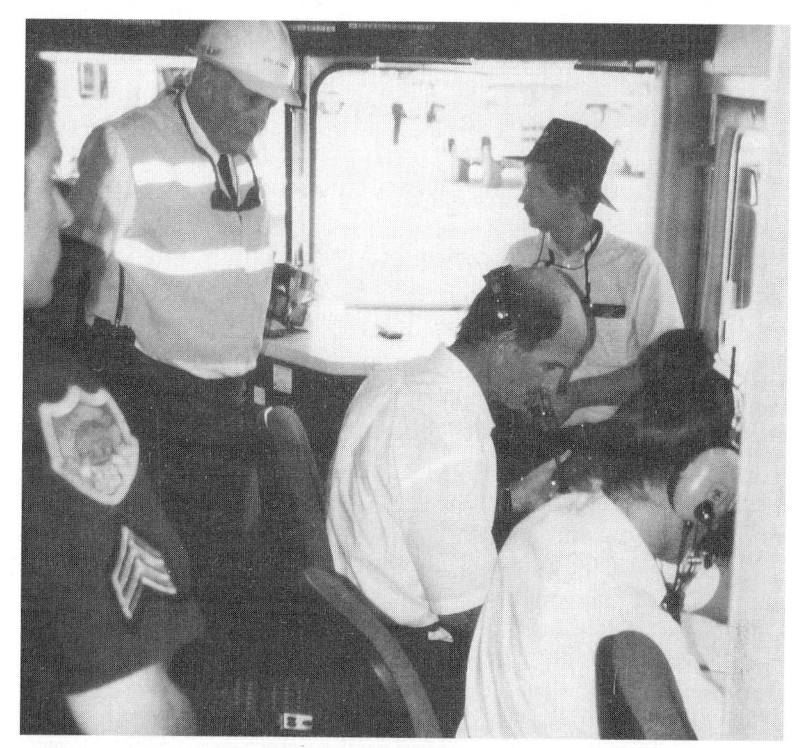

FIGURE 9-2 As a part of the command staff, the ISO must maintain communications with the incident commander and must listen to and understand messages to and from the operational components operating at an incident. Photo by Gerry Suftko, Mesa (AZ) Fire Department.

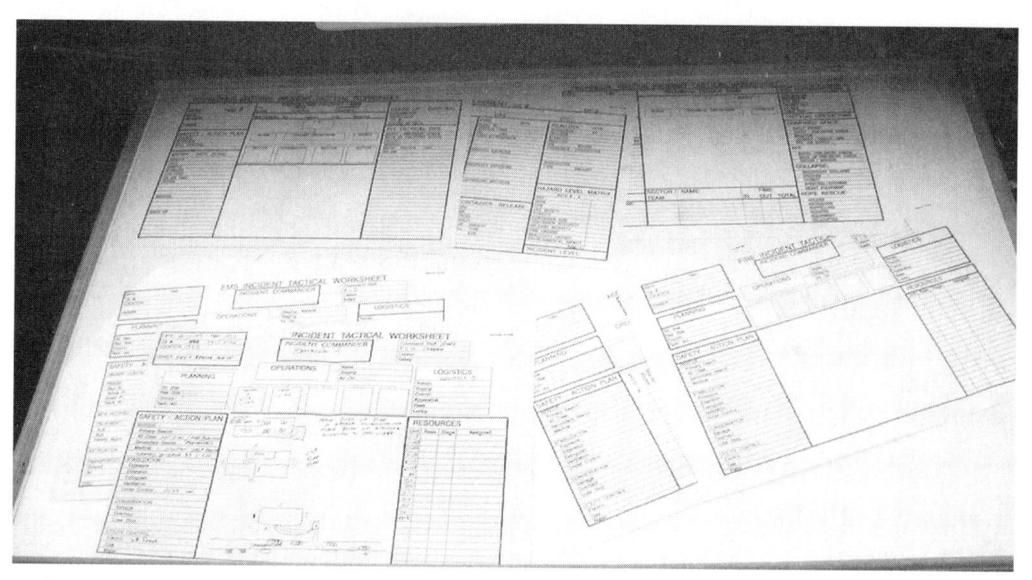

FIGURE 9-3 As operations are being terminated, the ISO and incident commander should meet face-to-face to brief each other on the positive and negative safety aspects of the incident. Documentation from the incident and information from the briefing should be discussed at the post-incident analysis. Photo courtesy Chesterfield County (VA) Fire and EMS.

The Monitoring Function at the Incident Scene

The IC and the ISO should develop an incident action plan relating to safety and risk management for each incident.

The IC develops the incident action plan based on the needs and severity of the incident. The incident action plan will incorporate safety and risk management issues, as well as strategy, tactics, and resources. An incident action plan for a company officer (CO) who is extinguishing a fully involved car fire is different from that of a CO or IC trying to extinguish a fully involved structure. With or without an ISO, the IC must monitor the incident. The priorities are always the same—*safety and risk management.*

The ISO must establish an incident safety plan at each incident. The monitoring function can be based on information gathered from the dispatcher, first-arriving companies, the IC, and personal observations. The ISO must monitor continuously throughout the incident by observation and communication (see Figure 9-4).

FIGURE 9-4 An incident action plan addresses the incident priorities, strategic goals, and tactical objectives of an incident, and incorporates safety and risk management issues. The ISO works to support this plan by monitoring the scene, and focusing on the safety and health of responders. Observation and communication are the keys to a successful ISO. Photo by Gerry Suftko, Mesa (AZ) Fire Department.

Risk Management

Risk management principles must be part of the IC's and ISO's incident action plan at each incident. All risk/benefit decisions are based on the emergency incident risk management guidelines discussed in chapter 8.

- Emergency responders should risk a lot only to save a lot.
- Emergency responders should risk only a little to save a little.
- Emergency responders should risk nothing to save what is already lost.

The IC will not have an ISO assigned at every incident. This means that the incident management system policy must reflect the need for continuous risk assessment. This will be a standard part of the IC's incident action plan.

The ISO must make risk assessment or analysis a primary function of the monitoring process. The skills and abilities gained through quality training, education, and experience as a firefighter, CO, and perhaps as a chief officer all serve as a framework to assist the ISO with risk assessment. Figure 9-5 cites one example of how such ISO training paid off at an incident.

Forecasting is a fundamental part of the risk management process. Forecasting is based on effective and safe department procedures or guidelines, experience

FIGURE 9-5 Responding to this vehicle fire, a battalion chief recognized the hazards presented by a burning recreational vehicle after having seen such a scenario in the National Fire Academy's course, Incident Safety Officer. He pulled all personnel back; no one was injured when the propane tanks "let go" just as in the course. Training, education, and experience are all important for an effective ISO. Photo by Mike Connolly, Marion County (FL) Fire-Rescue.

with similar incidents in the past, and monitoring conditions of personnel, the structure, and/or the environment at the current incident.

Responsibilities of the ISO

The ISO must have a defined role within the ICS. One of the most important is leading by example. If an ISO is appointed at a working fire and then walks around the incident scene with no PPE, that ISO sets a poor example for department personnel.

Monitoring unsafe acts, which cause the highest percentage of accidents, injuries, and fatalities, must be part of the incident action plan (see Figure 9-6). The ISO should watch for the following situations.

- Lack of use or improper use of PPE. The type of incident will dictate the proper PPE.
- Personnel accountability system not being used at the incident. Freelancing and lack of crew or company accountability is a major issue.
- Improper operations at the incident. For example, you observe defensive and offensive attacks being conducted at the same time.

Unsafe conditions also create situations that can lead to accidents and injuries at an incident scene. As the ISO monitors, he or she looks for conditions that could damage apparatus and equipment. The structural integrity of the building, tank, or container must be monitored, based on the length of the operation, fire con-

FIGURE 9-6 The ISO must monitor the scene for unsafe acts, which cause the highest percentage of accidents, injuries, and fatalities. Photo by Steve Weissman, Fairfax County (VA) Fire and Rescue Department.

ditions, or any other circumstance that could jeopardize personnel safety. The ISO shown in Figure 9-7 is assessing overall structural integrity prior to overhaul operations.

Fatigue, medical problems, mental health, or any other condition that is affecting a firefighter's physical or mental well-being must be addressed. If the ISO notes any problems, they must be relayed immediately to the IC for action.

Critical incident stress is a normal response experienced by normal persons following an abnormal event. It can be a result of trauma experienced by emergency services personnel. These reactions must not be perceived as weakness, mental infirmity, or any other aberration. Emotional stress following a significant incident is a very normal reaction. Although it cannot be prevented, it can be treated.

FIGURE 9-7 The ISO must monitor the scene for unsafe conditions that can lead to accidents and injuries, and must cause the appropriate preventative action to take place. Photo by Scott Boatwright, Fairfax County (VA) Fire and Rescue Department.

Critical incident stress management (CISM) provides an intervention process that will address these problems in a systematic manner. This process will allow personnel to lessen the impact of the incident and will return them gradually to mental well-being. It is imperative that departments establish a CISM team. The ISO and the IC have the responsibility to initiate the CISM process if the need arises. The monitoring process by the ISO will assist the IC in maintaining the health and welfare of all personnel involved at the incident (see Figure 9-8).

Emergency Authority for the ISO

The ICS policy must dictate that the ISO has the authority to halt or suspend an operation at an incident based on danger or hazards to personnel or equipment. This emergency authority must be supported by the fire chief. This is an element of the process that, it is hoped, will never have to be used. As long as the IC and the ISO are doing their jobs, the need never may arise. Exercising this emergency authority is, without a doubt, a significant career event.

Should the ISO observe a situation in which they must intervene to stop or alter an operation, communication with the IC is critical. Stopping or suspending an operation may affect other personnel. Good judgment and justification for

FIGURE 9-8 In addition to exposure to various toxins, emergency responders are exposed to significant emotional stresses as well. The ISO should recognize the signs indicating when personnel may need critical incident stress management. Photo by Steve Weissman, Fairfax County (VA) Fire and Rescue Department.

stopping or changing operations is crucial. The ISO must inform the IC why he or she is stopping an operation.

It may be as simple as removing a ventilation crew from the roof of a building, or as complex as trying to remove multiple crews conducting offensive operations in a building when defensive operations have started.

It must be emphasized that the ISO should only alter, terminate, or suspend an operation based on an imminent hazard to personnel. If the ISO recognizes a problem with tactics or identifies a potential (but not imminent) hazard, these should be brought to the IC's attention. The emergency authority granted to the ISO must not be viewed as a "power thing," but rather as a potentially life-saving tool.

Safety at the Incident Scene

Several key issues must be remembered and practiced. Some of the situations that arise at the incident can be handled directly by the ISO; other situations may require additional assistance; and some are handled after the incident is over.

Communications

The ISO is under the direction of the IC and must maintain communications with the IC. The primary focus is safety. If the IC has the ISO concentrating on anything else but safety, the process is ineffective. A good exchange of information in a positive environment will foster a good relationship between these two persons.

It is important—perhaps critical—to recognize that the ISO is not a "safety cop." Enforcement of SOPs is not the issue at an incident; keeping responders from getting injured or killed is. The ISO should *never* be in a position where other responders dread their presence for fear of getting "written up." The ISO must be the responders' advocate and look out for their well-being.

A simple example involves personal protective equipment. If the ISO sees a member without gloves on at an EMS call, the ISO should hand the member a pair of gloves or say, "Can I get you some gloves?" Or, in Figure 9-9, simply hand the firefighter the helmet on the ground beside him. Threatening disciplinary action, taking notes, or chastising the member is not appropriate. The same is true, for example, with a PASS device not turned on at a structure fire (turn it on for the firefighter) or a face shield not down during an extrication (put it down for the member or motion to them to put it down).

Monitoring for Unsafe Acts, Unsafe Conditions, and Unsafe Operations

By observing and listening, the ISO must continuously monitor personnel operating at the incident. This becomes most important during the overhaul state or the final stages of the incident, when personnel may not consider safety a top priority. The type of situation, the weather or environmental conditions, and

FIGURE 9-9 The ISO should never develop a reputation of being a safety cop. Instead of enforcing safety standards through fear of ridicule or discipline, the ISO should coach or assist members in being safe. For example, the ISO could hand a member a pair of gloves or eye protection if they are without it, or turn on their PASS device for them if they forget. Photo by Gordon M. Sachs.

apparatus and equipment placement are typical conditions that require the attention of the ISO. Figure 9-10 shows a typical yet often overlooked hazard on the fireground—traffic around water supply apparatus.

Recommendations to the HSO

The ISO must work closely with the HSO to solve problems encountered at an incident. The post-incident analysis may require the HSO to initiate changes in policy relating to standard operating procedures (SOPs).

If SOPs are not being followed at the incident because they cause unsafe acts, they may need to be revised. This may lead to development of new policy or revisions of current ones.

Training and education issues need to be addressed based on the outcome of an incident and/or the PIA. The HSO is in a position to provide the training or to ensure that it is done.

Incidents Involving Other Agencies

When the fire or EMS department has overall responsibility for the incident, as in Figure 9-11, the IC should advise the other agencies that the ISO is overseeing the

FIGURE 9-10 The ISO needs to monitor all aspects of an incident, including the "non-hazard" areas. Simple hazards (such as traffic—even if the apparatus is off the road) can prove fatal; a few cones could make this situation much safer for the firefighters in the picture. Photo by Robert Rosensteel, Sr., Vigilant Hose Company, Emmitsburg, MD.

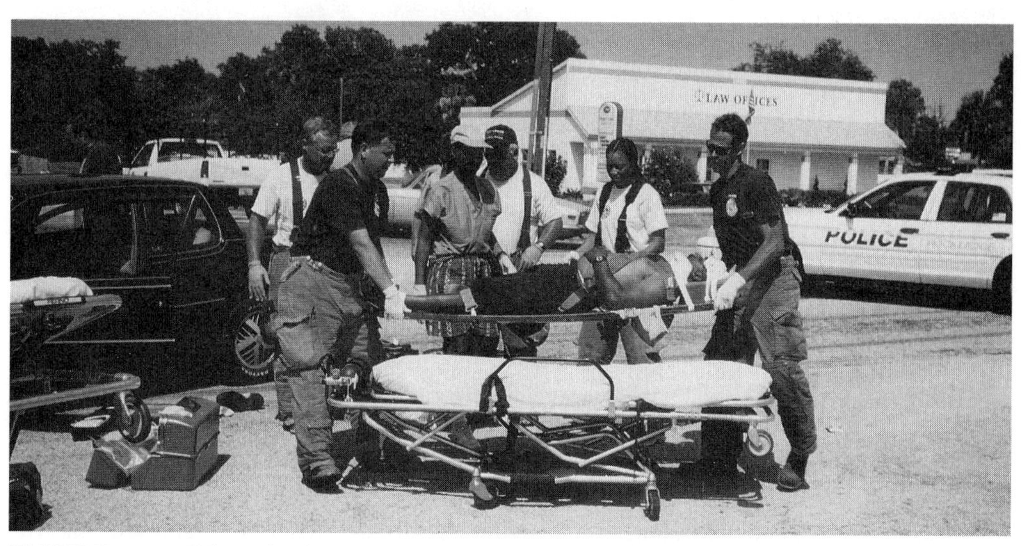

FIGURE 9-11 When the fire department is in charge of an incident, the fire department's ISO has the responsibility to monitor the safety of everyone (from all agencies) operating at the scene. Photo by Orlando Dominguez, Brevard County (FL) Fire/Rescue.

safety of everyone at the incident. The IC will make specific notification to all agencies involved with personnel present at the incident. (The liaison officer may be the individual who actually delivers this notification.)

If the fire or EMS department is not the primary agency, the department and the ISO are specifically responsible for safety of all department personnel. A relationship should be established with the safety officer for the primary agency by contacting the in-charge agency's ISO and asking if they need assistance (see Figure 9-12). If the primary agency has never heard of a safety officer, explain the role of the safety officer and offer to help out by providing one for the incident. Perhaps you can provide training and advice later so they can establish such a position.

The ISO's primary function is to ensure the safety of fire department personnel. If the agency in charge does not subscribe to that philosophy, fire department

FIGURE 9-12 When the fire department is *not* the lead agency at an incident, the fire department's ISO is responsible only for the safety of fire department personnel. The fire department's ISO should work closely with the overall ISO (from the lead agency); if the lead agency does not have an ISO, the fire department can offer to provide one for the incident. Photo by Gerry Suftko, Mesa (AZ) Fire Department.

personnel should be tactfully removed from participation in the incident until the matter is resolved.

Routine training and contacts with other public safety and governmental agencies can improve and enhance relations during an emergency incident. It is very difficult to solve these problems at an incident—especially an incident that has multiple agencies involved.

SUMMARY

This safety function is part of the command staff duties. At an incident scene, there is one primary purpose—*incident safety.* The ISO is as important as the IC from the standpoint of safety and risk management. The responsibilities given to the ISO are crucial to the successful outcome of the incident and to personnel safety. This function, coupled with an aggressive and complete ICS, will prove beneficial.

The importance of using the risk management concept cannot be overstated. Risk management is a primary responsibility of the IC and supervisory personnel.

FIGURE 9-13 Properly trained fire and EMS department safety officers—HSOs and ISOs—can help reduce the number of line-of-duty fatalities suffered by emergency response personnel. Photo by Jack Jordan, Phoenix (AZ) Fire Department.

As the ISO undertakes the safety function, that becomes a priority, as well. In the past, the fire service has done very poorly from this perspective. As risk management is used more and more in the fire service, we will come to appreciate it even more. The tangible benefit will be a reduction in the number and severity of accidents and injuries.

By appointing an ISO, the IC has provided another mechanism to ensure safe incident operations. The ISO can function as the eyes and ears of the IC, monitoring the entire incident scene and communicating conditions. If the function is not assigned or is given to an unskilled and untrained individual, the consequences could be tragic.

The fire service collectively can reduce the number of firefighter fatalities, injuries, accidents to equipment and apparatus, and occupational health exposures. The fire service's poor safety record is proof that incident scene safety is a target for improvement. An emphasis on safety by all department personnel can improve this record.

Appendices

Tools to Assist the Fire and EMS Department Safety Officer

A

Annotated Listing of Laws, Regulations, and Standards Affecting the Fire and EMS Department Safety Officer

Laws, regulations, and standards establish the legal framework for fire service operations. The potential impact of these various types of legal authorities depends on several factors. For example, the applicability of federal Occupational Safety and Health Administration (OSHA) regulations is determined by the character of the employer/employee relationship, the nature of the state occupational safety and health agency, and legal precedent. Because the content and interpretation of statutes and codes can change, the information provided here is intended solely as an introduction to a complex topic. Fire and EMS department managers need to review and understand applicable laws, regulations, and standards in their entirety before making decisions about SOPs.

Laws

Under our constitutional system of governance, no federal laws exist that specifically define the duties and powers of local fire and EMS departments. States may have different laws regarding the organization, duties, and powers of fire and EMS agencies within that state. In addition, cities and counties may have statutory provisions that further define emergency services operations within their jurisdictions. Fire and EMS department officers should understand and be able to cite all statutes that define their authority.

Important federal laws that apply to all fire and EMS departments within the United States are summarized in the following sections.

Americans with Disabilities Act

Title II of the Americans with Disabilities Act (ADA) covers all activities of state and local governments regardless of the entity's size and amount of federal funding it receives. The United States Department of Justice (DOJ) promulgates and enforces regulations under ADA. Title II requires that state and local governments give people with disabilities an equal opportunity to benefit from their programs, services, and activities. Considerations include access to jobs and "reasonable" accommodations, important factors in planning how operations will be conducted (i.e., writing SOPs). Fire and EMS departments incorporated as not-for-profit entities may be covered under either Title II or Title III of ADA; functional requirements are essentially the same.

People with disabilities, including HIV/AIDS, are entitled to receive the same level of medical care as other citizens. There have been several notable instances where medical providers, including EMS personnel, were sued for failing to provide the appropriate level of care to people with HIV/AIDS or other disabilities.

All fire and EMS department dispatch centers or public safety answering points (PSAPs) are required to equip every dispatcher console with teletypewriters (TTYs). People with disabilities use these devices to communicate via telephone. Call-takers are required to query "silent calls" with the TTYs to ensure that they are not actually calls from a disabled person.

ADA Title III, *Public Accommodations,* contains architectural requirements for new and altered buildings that can influence the conduct of fire and EMS department emergency operations. For example, Title III mandates changes in elevator function and stairwell pressurization during fire alarm activations in multistory buildings. Fire and EMS personnel must understand how building alterations can affect evacuation and rescue procedures for people with disabilities.

Title II may impact fire and EMS department SOPs in a variety of ways. A great deal of useful information regarding ADA can be found on the DOJ web site, including hotline numbers for questions (see appendix B).

Ryan White Comprehensive AIDS Resources Emergency Act of 1990

The Ryan White Act requires that hospitals notify personnel that may have been exposed to an infectious disease during emergency response. The law further requires that each fire and EMS department appoint a designated officer, typically called an infection control officer (ICO), to facilitate communications between medical facilities and emergency responders. With airborne pathogens like tuberculosis, the law requires medical facilities to contact the ICO within 48 hours after determining that a patient treated by responders is infected. When a bloodborne pathogen is involved, the medical facility must notify the emergency responder after a written request from the ICO.

In 1994, the Centers for Disease Control (CDC) published *CDC Guidelines on Implementation of Provisions of the Ryan White Comprehensive AIDS Resources Emergency Act Regarding Emergency Response Employees.* This supplementary material, which focuses primarily on infectious disease education and planning for emergency responders, may be helpful for developing departmental SOPs. Although the CDC guidelines are neither regulations nor standards, many of the recommendations were incorporated into OSHA regulations and industry standards regarding bloodborne pathogens (described later).

OSHA General Duty Clause (29 USC 654(a)(1))

The intent of this statute is to protect employees from workplace accidents and exposures by requiring employers to recognize and correct hazards. In the absence of a specific OSHA regulation addressing a workplace hazard, OSHA may use national consensus standards like those developed by the NFPA to determine whether the existence of a workplace hazard violates the General Duty Clause. Fire and EMS departments can meet the requirements of 29 USC 654(a)(1) by removing workplace hazards and providing emergency responders with the appropriate training, equipment, and procedures to safely operate in hazardous environments (e.g., emergency scenes).

Superfund Amendments and Reauthorization Act

The Superfund Amendments and Reauthorization Act (SARA) mandates certain critical aspects of hazardous materials preparation and response, including training for emergency responders, the creation of state emergency response commissions (SERCs) and local emergency planning committees (LEPCs), state and local government planning activities, and hazardous waste reporting. SARA also mandated that OSHA and the EPA promulgate regulations governing hazardous materials training, operations, and emergency response. (More information on these regulations is presented later.)

Supreme Court Precedents

The legal community recognizes the outcome of specific court cases, called *precedents,* as an authority for deciding future cases. Courts at every level use precedents to help interpret laws and decide cases. The United States Supreme Court is the ultimate authority on the constitutionality of laws. For example, two precedent-setting cases decided by the Supreme Court concern fire-cause determination and arson investigation. *Michigan v. Tyler*[1] and *Michigan v. Clifford*[2] upheld that fire department personnel may remain in a building and seize evidence of arson without a warrant for a reasonable amount of time after a fire has been extinguished. However, any investigation activities after a property is released to the owner by the fire department must be conducted with a valid search warrant. Failure to obtain a warrant violates the citizens' rights under the Fourth Amendment to the Constitution.

Regulations

Regulations are rules established by government agencies to implement statutory laws. The applicability of federal regulations to fire and EMS department operations varies according to several factors that were discussed previously (see chapter 3).

State and local regulations also influence the SOP development process. The federal regulations described here represent a small sample of those that may be pertinent to fire and EMS departments.

Occupational Noise Exposure (OSHA 29 CFR § 1910.95)

Requires employers to measure sound levels in the workplace, provide protective hearing equipment, develop a hearing conservation program whenever employee noise exposures exceed permissible levels, and maintain records on employee noise exposure levels.

Hazardous Waste Operations and Emergency Response (HAZWOPER) (OSHA 29 CFR § 1910.120)

Applies to all personnel involved in hazardous materials response, whether volunteer or career, in every state where OSHA standards apply. The regulation requires employers to develop a comprehensive program for hazmat response and establishes minimum safety standards. (EPA 40 CFR § 311 is functionally similar to 29 CFR 1910.120 and covers emergency responders in all states, regardless of their status under OSHA.)

Fire Brigades (OSHA 29 CFR § 1910.156)

Refers to a wide variety of firefighting issues such as agency mission statements, training requirements, and personal protective equipment. Principally directed at industrial fire and EMS departments, the regulation can apply to state, county, and municipal fire and EMS departments in some jurisdictions.

Personal Protective Equipment (OSHA 29 CFR § 1910.132–§ 1910.140)

Establishes general requirements for employers to provide, test, inspect, and maintain personal protective equipment (PPE) for employees exposed to workplace hazards. Employees must be trained on the proper use of such equipment, to include eye protection, face protection, head and extremity protection, protective clothing, respiratory protection, and protective shields and barriers. In addition, 29 CFR § 1910.134 requires that, when employees enter a hazardous area using respiratory protection, one or more similarly equipped employees must be standing by to provide accountability and assist in rescue, if needed. Specific requirements are listed for regular maintenance and testing of respiratory equipment, fit testing, and other requirements. (All of the basic requirements applicable to private sector employees apply to firefighters and EMS personnel. Additional requirements apply specifically to fire suppression and rescue operations.)

Sanitation Requirements (OSHA 29 CFR § 1910.141)

Ensures that fire stations and other fire/EMS department facilities provide clean and sanitary work environments free from health hazards. The regulation addresses fixed facility requirements including general housekeeping, waste disposal, pest control, water supplies, toilet facilities, showers, change rooms, clothes-drying facilities, eating and drinking areas, waste disposal containers, sanitary storage, and food handling.

Permit-Required Confined Spaces (OSHA 29 CFR § 1910.146)

Intended to protect personnel who enter "permit-required confined spaces," as these terms are specifically defined in the regulation. An employer is required to issue a written permit to employees before they are allowed to enter a permit-required space. The portion of this regulation most applicable to emergency services personnel is paragraph K, *Rescue and Emergency Services.* Under paragraph K, fire and EMS departments that may respond to a confined-space incident are required to provide personnel with the appropriate PPE, rescue equipment, and training to perform rescues from permit-required spaces. Paragraph K does not require emergency services personnel to complete a permit before entry is made into a confined space for rescue purposes; however, a permit would be required to enter the space for training purposes. Rescuers must have atmospheric monitoring and ventilation equipment, lifelines and harnesses, a mechanical hoist system, communications equipment, and lighting equipment.

Lock-Out/Tag-Out Requirements (OSHA 29 CFR § 1910.147)

Intended to prevent injury to employees caused by the unexpected startup of machines or equipment or the release of stored energy. The rule mandates that emergency services personnel use certain safety measures to prevent the unexpected release of energy or startup of equipment. Lock-out/tag-out procedures may be necessary when performing rescues involving heavy industrial equipment, elevators, or electrical rooms. This standard also requires employers to create an employee protection program that defines lock-out/tag-out procedures.

Occupational Exposure to Bloodborne Pathogens (OSHA 29 CFR § 1910.1030)

Provides for employee protection from exposure to bloodborne pathogens or other potentially infectious materials. The regulation requires that fire and EMS departments establish a comprehensive education and control program for personnel who may be exposed to bloodborne pathogens or infectious materials. The program must cover the following topics: training for emergency services personnel about the dangers of bloodborne pathogens; methods to dispose of contaminated materials; disposal processes for "sharps," contaminated instruments and infectious materials; documentation of rescue worker exposures to infectious materials; and post-exposure medical evaluations. The department is also required to provide all protective equipment necessary to protect employees from bloodborne pathogens. Hepatitis B vaccinations must be offered at no cost to personnel.

Hazard Communication (OSHA 29 CFR § 1910.1200)

Requires that all employers (1) evaluate hazardous materials imported into, produced by, or used in a workplace and (2) communicate the resulting information to employees through labels, material safety data sheets (MSDS), and specialized training. In addition, employers must notify and educate employees about hazardous materials locations to which they may have to respond. All employers must develop a hazard communication plan and share copies of the plan and their MSDSs with local emergency responders. OSHA's definition of hazardous chemicals and specified threshold amounts determine which chemicals must be reported in these plans.

Trench/Collapse Rescue Operations (OSHA 29 CFR § 1926.650–§ 1926.652)

Establishes operational and safety practices for rescue incidents involving trenches. This rule prohibits entry into trenches that are not properly shored, specifies that emergency services personnel wear a lifeline into trenches, and requires that fire/EMS departments provide training to emergency services personnel about the hazards of trench operations.

Discrimination Against Employees Under OSHA Act of 1970 (OSHA 29 CFR § 1977)

Prevents employers from discriminating against employees who exercise the right to file OSHA-related complaints or testify against an employer during an investigation. The rule creates a procedure for employees to file grievances if they feel they are victims of retaliation. Employees who refuse to comply with occupational safety and health procedures implemented by an employer in accordance with OSHA rules are not entitled to protections afforded by this act.

Compressed Gas Cylinder Guidelines (DOT 49 CFR § 178, subpart C)

Regulates the construction, testing, and maintenance of compressed gas cylinders, including those for self-contained breathing apparatus and medical oxygen. Although this is a Department of Transportation (DOT) regulation, it incorporates consensus guidelines from the Compressed Gas Association (CGA) for inspecting and testing compressed gas cylinders. SOPs developed as part of a department's SCBA preventive maintenance and inspection program must ensure that compressed gas cylinders are hydrostatically tested at prescribed intervals.

Consensus Standards

Consensus standards created by the National Fire Protection Association (NFPA) and other professional organizations are extremely important to the development of effective SOPs. A detailed discussion of standards can be found in chapter 3. The following standards excerpts represent a small sample of those applicable to fire and EMS departments. Managers and team members should become familiar with the most current editions of these standards, in their entirety, during the SOP development process.

NFPA 471, Recommended Practice for Responding to Hazardous Materials Incidents

NFPA 471 outlines minimum requirements and operating guidelines for all organizations that have responsibilities when responding to hazardous materials incidents. The recommended practice specifically covers planning methods, policies, and procedures for determining incident levels, using personal protective equipment, decontamination, incident safety, and communications. Other topics include the use of control zones, monitoring equipment, incident mitigation measures, and medical monitoring.

NFPA 1250, Recommended Practice in Emergency Service Organization Risk Management

NFPA 1250 establishes minimum criteria to develop, implement, or evaluate an emergency service organization, risk management programs for effective risk identification, control, and financing.

NFPA 1403, Standard on Live-Fire Training Evolutions

This standard sets forth a systematic method to prepare for and conduct training evolutions involving live fire. The standard applies to live-fire training in specially constructed "burn buildings," as well as acquired structures. Requirements are organized into five categories: Acquired Structures, Gas-Fired Training Center Buildings, Non-Gas-Fired Training Center Buildings, Exterior Props, and Exterior Class B Fires. Within each category, guidelines are specified for student prerequisites, structures and facilities, fuel materials, safety, and instructors. Requirements for record keeping and reporting are also identified.

NFPA 1404, Standard for a Fire Department Self-Contained Breathing Apparatus Program

NFPA 1404 establishes fire/EMS department guidelines for developing a preventive maintenance and training program for self-contained breathing apparatus (SCBA). The standard is designed to meet or exceed federal requirements for worker respiratory protection programs. It identifies minimum program requirements and safety procedures for addressing provision of SCBA, emergency scene use, SCBA training certification, safe operation, in-service inspection, equipment maintenance, breathing air quality control, and program evaluation.

NFPA 1410, Standard on Training for Initial Fire Attack

This standard contains minimum requirements for the evaluation of training in initial fire flow delivery procedures by fire department personnel engaged in structural firefighting. It serves as a standard mechanism for evaluating minimum acceptable performance for hose line and water supply activities during training for initial fire attack. The standard describes methods of evaluation and logistical considerations for basic evolutions that can be adapted to local conditions. Required performance guidelines are represented for handlines, master streams, and automatic sprinkler system support.

NFPA 1470, Standard on Search and Rescue Training for Structural Collapse Incidents

This standard identifies and establishes levels of training for safely and effectively conducting operations at structural collapse incidents. It is designed to help organizations assess the level of operational capability needed, and to establish training and safety criteria. Specific training requirements are defined for personnel at three levels: Basic Operations, Medium Operations, and Heavy Operations. In addition, general safety requirements are identified, including appointment of a safety officer, use of personal protective equipment, use of other safety equipment, incident management, and physical fitness of personnel.

NFPA 1500, Standard on Fire Department Occupational Safety and Health Program

NFPA 1500 establishes minimum standards for fire service occupational safety and health programs. It applies to all aspects of the workplace, including incident scene and nonemergency operations. This broad standard requires departments to develop a comprehensive written risk management plan and an occupational safety and health program; designate a safety and health officer; appoint a safety and health committee; use incident command, personnel accountability, and safety systems at incidents; establish written SOPs; and maintain a data collection system and permanent record of job-related accidents, injuries, illnesses, and exposures. It also requires that responders maintain minimum levels of health and fitness and use personal protective equipment.

NFPA 1521, Standard for Fire Department Safety Officer

This standard contains minimum requirements for the assignment, duties, and responsibilities of fire/EMS department health and safety officers and incident safety officers. Related organizational requirements are defined, including personnel assignments and backup capabilities. The qualifications and authority of both positions are also described. Functions of the health and safety officer are defined in relation to risk management, safety program rules and SOPs, training and education, accident prevention and investigation, records management and data analysis, apparatus and equipment, facility inspection, health maintenance, infection control, critical incident stress management, and post-incident analysis. Functions of the incident safety officer are also described, to include participation in the incident management system, incident scene safety, fire suppression, emergency medical services operations, hazardous materials operations, and special operations.

NFPA 1561, Standard on Fire Department Incident Management System

This standard establishes a generic structure for the coordination and management of emergency incidents to help ensure the health and safety of emergency responders. It requires the adoption of an incident management system for command and control of all emergency incidents and training exercises. Written plans should be created to anticipate incidents that require standardized procedures and mutual aid with other agencies involved in emergency incidents. Departments should create a command structure and standard supervisory assignments, including incident command, command staff, planning, logistics, operations, communications, staging, and finance functions. Departments are also required to implement a personnel accountability system and address rehabilitation for all members operating at an incident.

NFPA 1581, Standard on Fire Department Infection Control Program

This standard contains minimum requirements for programs to control infectious and communicable disease hazards in the fire department work environment. It is applicable to organizations that provide fire suppression, rescue, emergency medical care, and other emergency services, including public, military, private, and industrial fire departments. The standard identifies minimum criteria for infection control in the fire station, at an incident scene, and at any other area where fire department members are involved in routine or emergency operations. Departments are directed to develop a written infection control policy and risk management plan, to conduct annual training and education programs for all members, and to designate an infection control officer. Other topics include vaccination programs, exposure control techniques, facility/station safety, cleaning and disinfecting, disposal methods, emergency medical operations and equipment, housekeeping, and labeling.

NFPA 1582, Standard on Medical Requirements for Fire Fighters and Information for Fire Department Physicians

This standard contains medical requirements for members, including full-time or part-time employees and paid or unpaid volunteers. It also provides information for physicians regarding other areas of emergency services medicine, including infection control and emergency incident rehabilitation. The standard specifies the minimum medical requirements for candidates and current members of fire departments and emergency service organizations. It is intended to provide information that will assist the department physician in providing proper medical support for members, help ensure that candidates and current members will be medically capable of performing their required duties, and help to reduce the risk of occupational injuries and illnesses.

NFPA 1670, Standard on Operations and Training for Technical Rescue Incidents

This standard identifies and establishes levels of functional capability for safely and effectively conducting operations at technical rescue incidents. It applies to organizations and departments that provide technical rescue response. The standard states that the authority having jurisdiction must have standard operating procedures at the awareness, operations, or technician level and must have operational procedures in place to perform safely at technical rescue incidents. Additionally, the standard calls for incident response planning and the provision of appropriate rescue equipment and personal protective equipment. The standard also covers specific types of technical rescue incidents.

ENDNOTES

1. 436 U.S. 499 (1978)
2. 464 U.S. 287, 294, 104 S.Ct. 641 (1984)

B

Organizations with Resources Related to Fire and EMS Department Occupational Safety and Health

Resources

American College of Emergency Physicians
P.O. Box 619911
Dallas, TX 75261-9911
(800) 798-1822
(972) 580-2816 fax
www.acep.org

Associated Public-Safety Communications Officers International, Inc. (APCO)
2040 S. Ridgewood Avenue
South Daytona, FL 32119
(888) 272-6911
(904) 322-2501 fax
www.apcointl.org

ASTM
100 Barr Harbor Drive
West Conshohacken, PA 19428
(610) 832-9730
(610) 832-9666 fax
www.astm.org

Bureau of Alcohol, Tobacco and Firearms

Department of the Treasury
650 Massachusetts Avenue, NW
Washington, DC 20226
www.atf.treas.gov

Bureau of Justice Assistance

United States Department of Justice
810 Seventh Street, NW
Washington, DC 20531
(202) 616-6500
www.ojp.usdoj.gov/BJA

Centers for Disease Control and Prevention

1600 Clifton Road, NE
Atlanta, GA 30333
(404) 639-3311
www.cdc.gov

Congressional Fire Service Institute (CFSI)

900 Second Street, NE, Suite 303
Washington, DC 20002
(202) 371-1277
(202) 682-3473 fax
www.cfsi.org

Department of Justice—Disability Rights Section

Civil Rights Division
P.O. Box 66738
Washington, DC 20035-6738
ADA Information Line: (800) 514-0301
ADA Information Line (TDD): (800) 514-0383
www.usdoj.gov/crt/ada/adahom1.htm

Department of Justice—Drug Enforcement Agency

Information Services Section (CPI)
700 Army-Navy Drive
Arlington, VA 22202
www.usdoj.gov/dea/

Department of Health and Human Services

HFE-88
5600 Fishers Lane
Rockville, MD 20857
(800) 532-4440
www.fda.gov

Disaster Research Center

University of Delaware
77 East Main Street
Newark, DE 19716
(302) 831-6618
(302) 831-2091 fax
www.udel.edu/drc

Emergency Management Institute

Federal Emergency Management Agency
National Emergency Training Center
16825 South Seton Avenue
Emmitsburg, MD 21727
(800) 238-3358
(301) 447-1497 fax
www.fema.gov

Federal Bureau of Investigation

935 Pennsylvania Avenue, NW
Washington, DC 20535-0001
(202) 324-3000
www.fbi.gov

Federal Emergency Management Agency

500 C Street, SW
Washington, DC 20472
(202) 646-4600
www.fema.gov

Fire Department Safety Officers' Association

P.O. Box 149
Ashland, MA 01721-0149
(508) 881-3114
(508) 881-1128 fax
www.fdsoa.org

Food and Drug Administration

Department of Health and Human Services
5600 Fishers Lane
Rockville, MD 20857
(800) 532-4440
www.fda.gov

Health Care Financing Administration

Department of Health and Human Services
7500 Security Boulevard
Baltimore, Maryland 21244
(410) 786-3000
www.hcfa.gov

International Association of Arson Investigators

300 S. Broadway, Suite 100
St. Louis, MO 63102
(314) 621-1966
(314) 621-5125 fax
www.fire-investigators.org

International Association of Emergency Managers

111 Park Place
Falls Church, VA 22046-4513
(703) 538-1795
(703) 241-5603 fax
www.emassociation.org

International Association of Fire Chiefs

4025 Fair Ridge Dr.
Fairfax, VA 22033-2868
(703) 273-0911
(703) 273-9363 fax
www.iafc.org

International Association of Fire Fighters

1750 New York Avenue
Washington, D.C. 20006
(202) 737-8484
(202) 783-4570 fax
www.iaff.org

International Society of Fire Service Instructors

P.O. Box 2320
Stafford, VA 22555
(800) 435-0005
(540) 657-0154 fax
www.isfsi.org

International Association of Wildland Fire

P.O. Box 328
Fairfield, WA 99012
(509) 283-2397

International City/County Management Association

777 North Capitol Street, NE, Suite 500
Washington, DC 20002-4201
(202) 289-4262
(202) 962-3500 fax
www.icma.org

International Personnel Management Association (IPMA)

1617 Duke Street
Alexandria, VA 22314
(703) 549-7100
(703) 684-0948 fax
www.ipma-hr.org

National Association of Emergency Medical Service Physicians

P.O. Box 15945-281
Lenexa, KS 66285-5945
(800) 228-3677
(913) 541-0156 fax
www.naemsp.org

National Association of Emergency Medical Technicians

408 Monroe Street
Clinton, MS 39056-4210
(800) 34-NAEMT
www.naemt.org

National Association of Search and Rescue

4500 Southgate Place
Suite 100
Chantilly, VA 20151-1714
(703) 222-6277
(703) 222-6283 fax
www.nasar.org

National Association of State EMS Directors

111 Park Place
Falls Church, VA 22046
(703) 538-1799
(703) 241-5603 fax

National Emergency Management Association

Council of State Governments
P.O. Box 11910
Lexington, KY 40578
(606) 244-8000
(606) 244-8239 fax
www.nemaweb.org

National Fire Protection Association

1 Batterymarch Park
Quincy, MA 02269-9101
(617) 770-3000
(617) 770-0700 fax
www.nfpa.org

National Highway Transportation Safety Administration (NHTSA)

Emergency Medical Services Division
400 Seventh Street, SW, NTS-14
Washington, D.C. 20590
(202) 366-5440
(202) 366-7731 fax
www.nhtsa.dot.gov/people/injury/ems

National Institute of Occupational Safety and Health

200 Independence Avenue
Washington, DC 20201
(800) 35-NIOSH
www.cdc.gov/niosh

National Institute of Standards and Technology

Fire Research Information Services
Building and Fire Research Laboratory
Gaithersburg, MD 20899
(301) 975-NIST
www.bfrl.nist.gov

National Safety Council

1121 Spring Lake Drive
Itasca, IL 60143-3201
(630) 285-1121
(630) 285-1315 fax
www.nsc.org

National Volunteer Fire Council

1050 17th Street, NW
Suite 1212
Washington, DC 20036
(888) ASK-NVFC
(202) 887-5291 fax
www.nvfc.org

Occupational Safety and Health Administration

United States Department of Labor
200 Constitution Avenue
Washington, DC 20210
(202) 693-1999
www.osha.gov/index.html

Public Risk Management Association

1815 N. Fort Myer Drive
Suite 1020
Arlington, VA 22209
(703) 528-7701
(703) 528-7966 fax

Risk and Insurance Management Society

655 Third Avenue
New York, NY 10017-5637
(212) 286-9292
(212) 986-9716 fax

United States Fire Administration

Federal Emergency Management Agency
National Emergency Training Center
16825 South Seton Avenue
Emmitsburg, MD 21727
(800) 238-3358
(301) 447-1270 fax
www.usfa.fema.gov

VFIS, Inc.

P. O. Box 2726
183 Leader Heights Road
York, PA 17405
(800) 233-1957
(717) 741-3130 fax
www.vfis.com

Women in the Fire Service

P.O. Box 5446
Madison, WI 53705
(608) 233-4768
(608) 233-4879 fax

C

Sample Reports and Investigation Forms

Personal Injury/Illness Investigation Report
Vehicle Accident/Loss Investigation Report
Infectious Exposure Form
Incident Exposure Record

Courtesy VFIS, Inc.

PERSONAL INJURY/ILLNESS INVESTIGATION REPORT

Emergency Service Organization________________ Date________

Address________________

Name of Injured________________ Date of Birth________

Address of Injured________________

Phone(   )________ Age____ Sex____ Height______ Weight______

Occupation________________ Job Title________________

Social Security Number________________ Years with Dept.________

Date of Injury________ Time of Injury________

Date Reported________ Time Reported________

Accident Reported To________________

Nature of Injury

- ☐ Fractures
- ☐ Inflammation
- ☐ Infectious Disease
- ☐ Frostbite, Cold Exposure
- ☐ Pinched Nerve, Ruptured Disk
- ☐ Electric Shock
- ☐ Chemical Injury
- ☐ Multiple Injury
- ☐ Recurrence
- ☐ Strain, Sprain, Torn Ligament
- ☐ Cuts, Lacerations, Punctures
- ☐ Inhalation, Fumes
- ☐ Inhalation, Smoke
- ☐ Heat Exhaustion, Fatigue
- ☐ Abrasions, Contusions, Bruises
- ☐ Heart Malfunction
- ☐ Eye Injury
- ☐ Burns
- ☐ Other ________

Parts of Body Affected

- ☐ Multiple Parts
- ☐ Head
- ☐ Eye(s)
- ☐ Ear(s)
- ☐ Neck
- ☐ Shoulder
- ☐ Chest
- ☐ Lung
- ☐ Abdomen
- ☐ Back
- ☐ Heart
- ☐ Groin
- ☐ Arm
- ☐ Hand
- ☐ Finger
- ☐ Leg(s)
- ☐ Knee(s)
- ☐ Ankle(s)
- ☐ Foot/Feet
- ☐ Ribs
- ☐ Hip
- ☐ Other ________

Where Injury Occurred

- ☐ Station Maintenance
- ☐ Apparatus Maintenance
- ☐ Emergency Scene
- ☐ Private Auto to Emergency
- ☐ Private Auto Non-Emergency
- ☐ Fundraising
- ☐ Convention
- ☐ Emergency Vehicle to Emergency
- ☐ Emergency Vehicle Non-Emergency
- ☐ Parades, Picnics, Contests
- ☐ Standing By Station for Call
- ☐ Training
- ☐ Auxiliary Services
- ☐ Responding/Returning to Emergency (Non-Vehicle)
- ☐ Other ________

Cause of Injury

- ☐ Fall
- ☐ Weather
- ☐ Making Safety Devices Inoperative
- ☐ Using Defective Equipment
- ☐ Using Equipment Improperly
- ☐ Failure to Use Personal Protection Equipment
- ☐ Struck By Object
- ☐ Improper Lifting
- ☐ Horseplay
- ☐ Structural Collapse
- ☐ Inadequate Guards or Protection
- ☐ Back Draft
- ☐ Improper Placement
- ☐ Civil Disturbance
- ☐ Inadequate Illumination
- ☐ Inadequate Ventilation
- ☐ Lack of Knowledge or Skill
- ☐ Irrational Civilian
- ☐ Communication
- ☐ Abuse or Misuse
- ☐ Other ________

Injury Occurred - Performing What Task?

- ☐ Forcible Entry
- ☐ Using Ladders
- ☐ Advancing/Directing Hose Line
- ☐ Ventilating
- ☐ Overhauling
- ☐ Salvage
- ☐ Servicing/Repairing Equipment
- ☐ Extrication
- ☐ Rescue Operation
- ☐ Administering Medical Aid
- ☐ Physical Fitness
- ☐ Other ________

Witness(es) to Injury: ________________

Injured Person's Signature________________ Date________

INVESTIGATION REPORT

Thoroughly describe accident: (What, How, Where, Equipment, Activity, etc.) ______

Hospitalized or Treated, Where? (Include Address) ______

Name and Address of Physician: (Include Referral) ______

Did the injury require individual to perform limited duties, or to be assigned to other duties or positions? YES or NO If yes, what duties or position? ______

And, what period of time? ______

Investigated by ______ Title ______ Date ______

SAFETY OFFICER'S REPORT:

What Acts, Failures to Act and/or Conditions Contributed Most Directly to This Accident? (Immediate Cause)

What Are the Basic or Fundamental Reasons for the Existence of These Acts and/or Conditions? (Fundamental Cause)

What Action Has or Will Be Taken to Prevent Recurrence? Place "X" By Items Completed.

Reviewed by Safety Officer ______ Title ______ Date ______

WP\WINDORD\VFIS\FORMS\PIREPORT.DOC - 2/96

C10.086

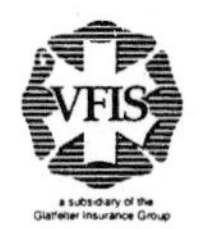

Vehicle Accident/Loss Investigation Report

(This is not a claim form)

Fire Department ____________________ Date ____________

Address ____________________

Name of Driver ____________________ Vehicle ID/Unit Number ____________

Type of Vehicle ____________________

Date Driver Last Certified On Above Vehicle ____________________

Date of Accident ____________ Time ____________ Date Reported ____________

Location of Accident ____________________

Roadway

- ☐ Straight
- ☐ Curve
- ☐ On Grade
- ☐ Level
- ☐ Hillcrest
- ☐ Dry
- ☐ Wet
- ☐ Muddy
- ☐ Snowy
- ☐ Icy
- ☐ Oily

- ☐ 2-lane
- ☐ 3-lane
- ☐ 4-lane
- ☐ Divided
- ☐ Rural
- ☐ Other ________
- ☐ Lanes marked
- ☐ Lanes unmarked
- ☐ No road defects
- ☐ Holes, ruts, etc.
- ☐ Loose material
- ☐ Other ________

Accident Occurred:

- ☐ At station
- ☐ Responding to emergency
- ☐ At emergency scene
- ☐ Returning from emergency
- ☐ Training
- ☐ Convention or parade
- ☐ Other ________

Type of Loss

- ☐ Personal injury
- ☐ Property damage
- ☐ Vehicle damage

Weather

- ☐ Clear
- ☐ Rain
- ☐ Snow
- ☐ Sleet
- ☐ Fog
- ☐ Other ________

Description Of Accident ____________________

____

____

____

____

____

Motor Vehicle Diagram

Complete the following diagram showing direction and positions of automobiles involved, designating clearly point of contact.

Indicate North ↑

Give Street Names and Directions

Your Vehicle

Other Vehicle 1 2

Instructions:

1. Show vehicles and direction of travel
2. Use solid line to show path of each vehicle before accident dotted line after accident...

-over-

Safety Analysis

What acts, failures to act and/or conditions contributed most directly to this accident? (Immediate Cause)

What are the basic or fundamental reasons for the existence of these acts and/or conditions? (Fundamental Cause)

What action has or will be taken to prevent recurrence? Place "X" by items completed.

Safety Supervisor's Comments

Driver's Signature ____ Date ____

Supervisor's Signature ____ Date ____

Safety Supervisor's Signature ____ Date ____

Infectious Exposure Form

Exposed Member's Name: ______________________ Position: ______________

Soc. Sec. #: ______________________ Home Phone: ______________

Field Inc. #: ____________ Shift: __________ Company: ______________

Name of Patient: ______________________ Sex: __________

Age: ________ Address: ______________________

Suspected or Confirmed Disease: ______________________

Transported to: ______________________

Transported by: ______________________

Date of Exposure: ______________________ Time of Exposure: ______________

Type of Incident (auto accident, trauma): ______________________

Type of protective equipment utilized: ______________________

What where you exposed to:

Blood ______ Tears ______ Feces ______ Urine ______ Saliva ______

Vomitus ______ Sputum ______ Sweat ______ Other ______

What part(s) of your body became exposed? Be specific: ______________________

____

____

____

____

Did you have any open cuts, sores, or rashes that became exposed? Be specific: ______________

____

____

How did exposure occur? Be specific: ______________________

____

____

Did you seek medical attention? ______Yes ______No

Where? ______________________ Date: ______________

Contact Infection Control Supervisor: Date______________ Time: ______________

Supervisor's Signature: ______________________ Date: ______________

Member's Signature: ______________________ Date: ______________

Infection Control Supervisor's Report

Medical facility notified? Yes _______ No _______

If Yes:

Name of Facility: ______________________________ Date: ______________

Address of Facility: ______________________________

Name of Facility Contact: ______________________________

Confirmed Exposure: ______________________________

Member notified? Yes _______ No _______

Member's Signature: ______________________________ Date: ______________

Medical Follow-Up Action:

____

____

____

____

____

Remarks:

____

____

____

____

____

Infection Control Supervisor's Signature: ____________________ Date: ______________

Printed in the USA

C10:102 - 2/93

Incident Exposure Record

Name ______________________________

Date of Birth ______________ Social Security Number ______________

Incident Number ______________ Incident Date ______________

Officer In Charge ______________________________

Location Of Incident ______________________________

Description Of Incident ______________________________

____

____

____

Type Of Exposure: Inhalation ______________________________

Direct Contact ______________________________

Ingestion ______________________________

Materials Exposed To ______________________________

____

Type Of Decontamination ______________________________

Length Of Exposure (time) ______________________________

Symptoms (if any) ______________________________

____

Treatment At Scene ______________________________

____

Name Of Medical Facility ______________________________

Treatment Rendered ______________________________

Protective Clothing and Equipment Used During Incident (list) ______________

____

____

Additional Information ______________________________

____

____

Firefighter/EMS Signature ______________ Date ______________

Chief's Signature ______________ Date ______________

-over-

Safety Officer's Analysis

What acts, failures to act and/or conditions contributed most directly to this accident? (Immediate Cause)

___

___

___

___

What are the basic or fundamental reasons for the existence of these acts and/or conditions? (Fundamental Cause)

___

___

___

___

What action has or will be taken to prevent recurrence? Place "X" by items completed.

___

___

___

___

___

___

___

Safety Officer's Comments___

___

___

___

___

___

___

Safety Officer's Signature___ Date ___

Printed in USA C10: 087 8/94

D

Emergency Incident Rehabilitation

Courtesy FEMA/U.S. Fire Administration

Emergency Incident Rehabilitation

The physical and mental demands associated with firefighting and other emergency operations, coupled with the environmental dangers of extreme heat and humidity or extreme cold, create conditions that can have an adverse impact upon the safety and health of the individual emergency responder. Members who are not provided adequate rest and rehydration during emergency operations or training exercises are at increased risk for illness or injury, and may jeopardize the safety of others on the incident scene. When emergency responders become fatigued, their ability to operate safely is impared. As a result, their reaction time is reduced and their ability to make critical decisions diminishes. Rehabilitation is an essential element on the incident scene to prevent more serious conditions such as heat exhaustion or heat stroke from occuring.

The need for emergency incident rehabilitation is cited in several national standards. Recent studies have concluded that a properly implemented fireground rehabilitation program will result in fewer accidents and injuries to firefighters.[1] Moreover, responders who are given prompt and adequate time to rest and rehydrate may safely reenter the operational scene, which may reduce the requirement for additional staffing at an incident.

An emergency incident rehabilitation program can be established in any department with a minimal impact on human, fiscal, and equipment-related resources. A successful rehabilitation program will improve the morale of the department and increase the level of productivity. It fits into the framework of the incident management systems (also known as incident command systems or ICS) used by fire departments, emergency medical services, hazardous materials response teams, and special rescue teams across the country.

The United States Fire Administration (USFA), in an effort to reduce the incidence of emergency responder injury and death, has developed this sample Emergency Incident Rehabilitation Standard Operating Procedure (SOP). This SOP outlines the responsibilities of incident commanders, supervisors, and personnel; identifies the components of Rehabilitation Area establishment; and provides Rehabilitation guidelines. A sample Emergency Incident Rehabilitation Report form is also included.

The USFA acknowledges the efforts of the individuals, departments, and organizations who provided information for the development of this sample SOP, and who reviewed it for accuracy and clarity. Additional copies of this publication may be obtained by writing

USFA Publications,
Post Office Box 70274
Washington, DC, 20024

Source: United States Fire Administration, Federal Emergency Management Agency, *Emergency Incident Rehabilitation,* FA114, (Washington, D.C.: USFA Publications, 1992).

SAMPLE

Standard Operating Procedure (SOP)

EMERGENCY INCIDENT REHABILITATION

1. **PURPOSE.**

To ensure that the physical and mental condition of members operating at the scene of an emergency or a training exercise does not deteriorate to a point that affects the safety of each member or that jeopardizes the safety and integrity of the operation.

2. **SCOPE.**

This procedure shall apply to all emergency operations and training exercises where strenuous physical activity or exposure to heat or cold exist.

3. **RESPONSIBILITIES.**

a. *Incident Commander.*

The Incident Commander shall consider the circumstances of each incident and make adequate provisions early in the incident for the rest and rehabilitation for all members operating at the scene. These provi sions shall include: medical evaluation, treatment and monitoring; food and fluid replenishment; mental rest; and relief from extreme climatic conditions and the other environmental parameters of the incident. The rehabilitation shall include the provision of Emergency Medical Services (EMS) at the Basic Life Support (BLS) level or higher.

b. *Supervisors.*

All supervisors shall maintain an awareness of the condition of each member operating within their span of control and ensure that adequate steps are taken to provide for each member's safety and health. The command structure shall be utilized to request relief and the reassign ment of fatigued crews.

c. *Personnel.*

During periods of hot weather, members shall be encouraged to drink water and activity beverages throughout the work day. During any emergency incident or training evolution, all members shall advise their supervisor when they believe that their level of fatigue or exposure to heat or cold is approaching a level that could affect themselves, their crew, or the operation in which they are involved. Members shall also remain aware of the health and safety of other members of their crew.

4. **ESTABLISHMENT OF REHABILITATION SECTOR.**

a. *Responsibility.*

The Incident Commander will establish a Rehabilitation Sector or Group when conditions indicate that rest and rehabilitation is needed for person nel operating at an incident scene or training evolution. A member will be placed in charge of the sector/group and shall be known as the Rehab Officer. The Rehab Officer will typically report to the Logistics Officer in the framework of the incident management system.

b. *Location.*

The location for the Rehabilitation Area will normally be designated by the Incident Commander. If a specific location has not been designated, the Rehab Officer shall select an appropriate location based on the site characteristics and designations below.

c. *Site Characteristics.*

(1) It should be in a location that will provide physical rest by allowing the body to recuperate from the demands and hazards of the emergency operation or training evolution.
(2) It should be far enough away from the scene that members may safely remove their turnout gear and SCBA and be afforded mental rest from the stress and pressure of the emergency operation or training evolution.[2]
(3) It should provide suitable protection from the prevailing environmental conditions. During hot weather, it should be in a cool, shaded area. During cold weather, it should be in a warm, dry area.
(4) It should enable members to be free of exhaust fumes from apparatus, vehicles, or equipment (including those involved in the Rehabilitation Sector/Group operations).
(5) It should be large enough to accomodate multiple crews, based on the size of the incident.
(6) It should be easily accessible by EMS units.
(7) It should allow prompt reentry back into the emergency operation upon complete recuperation.

d. *Site Designations.*

(1) A nearby garage, building lobby, or other structure.
(2) Several floors below a fire in a high rise building.
(3) A school bus, municipal bus, or bookmobile.
(4) Fire apparatus, ambulance, or other emergency vehicles at the scene or called to the scene.
(5) Retired fire apparatus or surplus government vehicle that has been renovated as a Rehabilitation Unit. (This unit could respond by request or be dispatched during certain weather conditions.)
(6) An open area in which a rehab Area can be created using tarps, fans, etc.

e. *Resources.*

The Rehab Officer shall secure all necessary resources required to adequately staff and supply the Rehabilitation Area. The supplies should include the items listed below:

(1) Fluids - water, activity beverage, oral electrolyte solutions and ice.
(2) Food - soup, broth, or stew in hot/cold cups.
(3) Medical - blood pressure cuffs, stethoscopes, oxygen administration devices, cardiac monitors, intravenous solutions and thermometers.[3]
(4) Other - awnings, fans, tarps, smoke ejectors, heaters, dry clothing, extra equipment, floodlights, blankets and towels, traffic cones and fireline tape (to identify the entrance and exit of the Rehabilitation Area).

5. **GUIDELINES.**

a. ***Rehabilitation Sector/Group Establishment.***

Rehabilitation should be considered by staff officers during the initial planning stages of an emergency response. However, the climatic or environmental conditions of the emergency scene should not be the sole justification for establishing a Rehabilitation Area. Any activity/ incident that is large in size, long in duration, and/or labor intensive will rapidly deplete the energy and strength of personnel and therefore merits consideration for rehabilitation.

Climatic or environmental conditions that indicate the need to establish a Rehabilitation Area are a heat stress index above 90 F (see table 1-1)[4] or windchill index below 10 F (see table 1-2).[5]

b. *Hydration.*

A critical factor in the prevention of heat injury is the maintenance of water and electrolytes. Water must be replaced during exercise periods and at emergency incidents. During heat stress, the member should consume at least one quart of water per hour. The rehydration solution should be a 50/50 mixture of water and a commercially prepared activity beverage and administered at about 40 F.[6] Rehydration is important even during cold weather operations where, despite the outside temperature, heat stress may occur during firefighting or other strenuous activity when protective equipment is worn. Alcohol and caffeine beverages should be avoided before and during heat stress because both interfere with the body's water conservation mechanisms.[7] Carbonated beverages should also be avoided.

c. *Nourishment.*

The department shall provide food at the scene of an extended incident when units are engaged for three or more hours. A cup of soup, broth, or stew is highly recommended because it is digested much faster than sandwiches and fastfood products. In addition, foods such as apples, oranges, and bananas provide supplemental forms of energy replacement. Fatty and/or salty foods should be avoided.

d. *Rest.*

The "two air bottle rule," or 45 minutes of worktime, is recommended as an acceptable level prior to mandatory rehabilitation. Members shall rehydrate (at least eight ounces) while SCBA cylinders are being changed. Firefighters having worked for two full 30-minute rated bottles, or 45 minutes, shall be immediately placed in the Rehabilitation Area for rest and evaluation. In all cases, the objective evaluation of a member's fatigue level shall be the criteria for rehab time. Rest shall not be less than ten minutes and may exceed an hour as determined by the Rehab Officer.[8] Fresh crews, or crews released from the Rehabilitation Sector/Group, shall be available in the Staging Area to ensure that fatigued members are not required to return to duty before they are rested, evaluated, and released by the Rehab Officer.

e. *Recovery.*

Members in the Rehabilitation Area should maintain a high level of hydration. Members should not be moved from a hot environment directly into an air conditioned area because the body's cooling system can shut down in response to the external cooling. An air conditioned environment is acceptable after a cool-down period at ambient temperature with sufficient air movement. Certain drugs impair the body's ability to sweat and extreme caution must be exercised if the member has taken antihistamines, such as Actifed or Benadryl, or has taken diuretics or stimulants.

f. *Medical Evaluation.*

(1) Emergency Medical Services (EMS) - EMS should be provided and staffed by the most highly trained and qualified EMS personnel on the scene (at a minimum of BLS level). They shall evaluate vital signs, examine members, and make proper disposition (return to duty, continued rehabilitation, or medical treatment and transport to medical facility). Continued rehabilitation should consist of additional monitoring of vital signs, providing rest, and providing fluids for rehydration. Medical treatment for members whose signs and/or symptoms indicate potential problems, should be provided in accordance with local medical control procedures. EMS personnel shall be assertive in an effort to find potential medical problems early.

(2) Heart Rate and Temperature-The heart rate should be measured for 30 seconds as early as possible in the rest period. If a member's heart rate exceeds 110 beats per minute, an oral temperature should be taken. If the member's temperature exceeds 100.6 F, he/she should not be permitted to wear protective equipment. If it is below 100.6 F and the heart rate remains above 110 beats per minute, rehabilitation time should be increased. If the heart rate is less than 110 beats per minute, the chance of heat stress is negligible.[9]

(3) Documentation-All medical evaluations shall be recorded on standard forms along with the member's name and complaints and must be signed, dated and timed by the Rehab Officer or his/her designee.

g. *Accountability.*

Members assigned to the Rehabilitation Sector/Group shall enter and exit the Rehabilitation Area as a crew. The crew designation, number of crew members, and the times of entry to and exit from the Rehabilitation Area shall be documented by the Rehab Officer or his/her designee on the Company Check-In/Out Sheet. Crews shall not leave the Rehabilitation Area until authorized to do so by the Rehab Officer.

HEAT STRESS INDEX

TEMPERATURE °F	RELATIVE HUMIDITY 10%	20%	30%	40%	50%	60%	70%	80%	90%
104	98	104	110	120	132				
102	97	101	108	117	125				
100	95	99	105	110	120	132			
98	93	97	101	106	110	125			
96	91	95	98	104	108	120	128		
94	89	93	95	100	105	111	122		
92	87	90	92	96	100	106	115	122	
90	85	88	90	92	96	100	106	114	122
88	82	86	87	89	93	95	100	106	115
86	80	84	85	87	90	92	96	100	109
84	78	81	83	85	86	89	91	95	99
82	77	79	80	81	84	86	89	91	95
80	75	77	78	79	81	83	85	86	89
78	72	75	77	78	79	80	81	83	85
76	70	72	75	76	77	77	77	78	79
74	68	70	73	74	75	75	75	76	77

NOTE: Add 10°F when protective clothing is worn and add 10°F when in direct sunlight.

HUMITURE °F	DANGER CATEGORY	INJURY THREAT
BELOW 60°	NONE	LITTLE OR NO DANGER UNDER NORMAL CIRCUMSTANCES
80° - 90°	CAUTION	FATIGUE POSSIBLE IF EXPOSURE IS PROLONGED AND THERE IS PHYSICAL ACTIVITY
90° - 105°	EXTREME CAUTION	HEAT CRAMPS AND HEAT EXHAUSTION POSSIBLE IF EXPOSURE IS PROLONGED AND THERE IS PHYSICAL ACTIVITY
105° - 130°	DANGER	HEAT CRAMPS OR EXHAUSTION LIKELY, HEAT STROKE POSSIBLE IF EXPOSURE IS PROLONGED AND THERE IS PHYSICAL ACTIVITY
ABOVE 130°	EXTREME DANGER	HEAT STROKE IMMINENT!

Table 1-1

WIND CHILL INDEX

WIND SPEED (MPH)	TEMPERATURE °F 45	40	35	30	25	20	15	10	5	0	-5	-10	-15
5	43	37	32	27	22	16	11	6	0	-5	-10	-15	-21
10	34	28	22	16	10	3	-3	-9	-15	-22	-27	-34	-40
15	29	23	16	9	2	-5	-11	-18	-25	-31	-38	- 45	-51
20	26	19	12	4	-3	-10	-17	-24	-31	-39	-46	-53	-60
25	23	16	8	1	-7	-15	-22	-29	-36	-44	-51	-59	-66
30	21	13	6	-2	-10	-18	-25	-33	-41	-49	-56	-64	-71
35	20	12	4	-4	-12	-20	-27	-35	-43	-52	-58	-67	-75
40	19	11	3	-5	-13	-21	-29	-37	-45	-53	-60	-69	-76
45	18	10	2	-6	-14	-22	-30	-38	-46	-54	-62	-70	-78

A B C

	WIND CHILL TEMPERATURE °F	DANGER
A	ABOVE -25° F	LITTLE DANGER FOR PROPERLY CLOTHED PERSON
B	-25° F / -75° F	INCREASING DANGER, FLESH MAY FREEZE
C	BELOW -75° F	GREAT DANGER, FLESH MAY FREEZE IN 30 SECONDS

Table 1-2

REHAB SECTOR COMPANY CHECK-IN / OUT SHEET[10]

CREWS OPERATING ON THE SCENE: ________________________________

____

UNIT #	# PERSONS	TIME IN	TIME OUT		UNIT #	# PERSONS	TIME IN	TIME OUT

EMERGENCY INCIDENT REHABILITATION REPORT[11]

INCIDENT: ____________________

DATE: ____________________

NAME / UNIT#	TIME(S)	TIME /# Bottles	BP	PULSE	RESP	TEMP	SKIN	TAKEN BY	COMPLAINTS/CONDITION	TRANSPORT?

Footnotes

[1] NIOSH, "Health Hazard Evaluation Report," HETA 90-395-2121. June 1991. pp. 7-9.

[2] "Time Out for Rehab," REKINDLE. July 1988, Vol. 1; Issue 7, pp. 15-16. Author Unknown.

[3] Carafano, Peter, "Firefighter Rehabilitation," FIRE CHIEF. March 1990. pp. 42-44.

[4] Sachs, Gordon, "Heat Related Stress," FAIRFAX COUNTY FIRE & RESCUE DEPARTMENT, SAFETY BULLETIN SB-87-08. June 1987. Attachment A.

[5] Sachs, Gordon, "Cold Exposure Windchill," FAIRFAX COUNTY FIRE & RESCUE DEPARTMENT, SAFETY BULLETIN SB-88-01. January 1988. Attachment A.

[6] Rose, Larry, "Drink and Thrive a Study of On-scene Rehabilitation," STRATEGIC ANALYSIS OF FIRE DEPARTMENT OPERA TIONS. September 1990. pp. 1-16.

[7] IAFF Department of Research, Health and Safety Division, "Ther mal Stress and the Firefighter", ISBN 0-942920-04-X. 1982. pp.21-22.

[8] Dodson, Dave, "Parker, Colorado Fire District Rehabilitation Plan," HEALTH AND SAFETY. December 1990.

[9] Skinner, James, "Coping with heat stress on the fireground. Fight ing the Fire Within," FIREHOUSE. August 1985. pp. 46-48 and 66.

[10] Becker, David S. and Goodson, Fred, "EMS-Rehab Sector Resource Manual," CHESTERFIELD FIRE PROTECTION DISTRICT. 1991. p. 5.

[11] Becker, David S. and Goodson, Fred, "EMS-Rehab Sector Resource Manual," CHESTERFIELD FIRE PROTECTION DISTRICT. 1991. p. 6.

E

Incident Safety Officer Checklists

Fire
MCI
Hazmat
Technical Rescue

INCIDENT SAFETY OFFICER CHECKLIST

FIRE INCIDENTS

_______ Incident safety officer wearing appropriate protective clothing/equipment, including command identification vest.

_______ Accountability tag given to incident commander.

_______ Face-to-face briefing with incident commander.

_______ Understand the incident commander's incident action plan.

_______ Ensure suitable, safe command post is set up and visible.

_______ Develop and implement an incident safety plan.

_______ Perform 360-degree walkaround.

_______ Ensure appropriate use of protective clothing/equipment by all members.

_______ Ensure personnel accountability system is being utilized appropriately.

_______ Conduct rapid emergency incident risk management analysis.

    _______ Risk a lot only to save a lot.

    _______ Risk a little only to save a little.

    _______ Risk nothing to save what is already lost.

_______ Ensure appropriate safety zones are set up.

_______ Ensure rapid intervention crew is in place as soon as practical.

_______ Ensure provision for rehab has been made (rehab area established).

_______ Ensure transport unit is committed at the scene.

_______ Consider the need for additional incident safety officers.

Notes/comments or PIA issues (positive and negative):

FIRE INCIDENTS

_______ Ensure that all personnel know the level of operation.

_______ Offensive

_______ Defensive (time: _______)

_______ Monitor fire conditions.

_______ Increasing

_______ Decreasing

_______ Monitor structural conditions.

_______ Identify building construction indicators.

_______ Establish additional incident safety officers as needed for scope of incident.

_______ Brief the rapid intervention crew.

_______ Ensure that all personnel are in crews (NO FREELANCING).

_______ Ensure roof operations are supervised.

_______ Ensure interior and roof crews have multiple means of egress.

_______ Ensure utilities are secured.

_______ Ensure crews are being rehabed.

_______ Have air quality monitored prior to SCBA removal (level: _______).

Other issues not covered (be specific):

____

____

____

Benchmarks (minutes)

10 20 30 40 50 60 70

INCIDENT SAFETY OFFICER CHECKLIST

MULTIPLE CASUALTY INCIDENTS

_______ Incident safety officer wearing appropriate protective clothing/equipment, including command identification vest.

_______ Accountability tag given to incident commander.

_______ Face-to-face briefing with incident commander.

_______ Understand the incident commander's incident action plan.

_______ Ensure suitable, safe command post is set up and visible.

_______ Develop and implement an incident safety plan.

_______ Perform 360-degree walkaround.

_______ Ensure appropriate use of protective clothing/equipment by all members.

_______ Ensure personnel accountability system is being utilized appropriately.

_______ Conduct rapid risk assessment/risk benefit analysis.

- _______ Risk a lot only to save a lot.
- _______ Risk a little only to save a little.
- _______ Risk nothing to save what is already lost.

_______ Ensure appropriate safety zones are set up.

_______ Ensure rapid intervention crew is in place as soon as practical.

_______ Ensure provision for rehab has been made (rehab area established).

_______ Ensure transport unit is committed at the scene.

_______ Consider the need for additional incident safety officers.

Notes/comments or PIA issues (positive and negative):

MULTIPLE CASUALTY INCIDENTS

_______ Scene safety and security is being addressed.

_______ Monitor use of appropriate personal protective clothing/equipment.

_______ Ensure proper use of EMS incident management system positions.

_______ Ensure utilities are secured.

_______ Ensure protective hoseline is in place (if applicable).

_______ Consider infection/exposure control.

    _______ Minimize exposure potential.

    _______ Personal protective equipment.

    _______ Decontamination.

_______ Establish additional incident safety officers as needed for scope of incident.

_______ Brief the rapid intervention crew.

_______ Ensure that all personnel are in crews (NO FREELANCING).

_______ Monitor landing zone safety or brief LZ officer.

_______ Ensure crews are being rehabed.

_______ Consider critical incident stress management (CISM).

Other issues not covered (be specific):

    ____

    ____

    ____

Benchmarks (minutes)

10   20   30   40   50   60   70

INCIDENT SAFETY OFFICER CHECKLIST

HAZARDOUS MATERIALS INCIDENTS

_______ Incident safety officer wearing appropriate protective clothing/equipment, including command identification vest.

_______ Accountability tag given to incident commander.

_______ Face-to-face briefing with incident commander.

_______ Understand the incident commander's incident action plan.

_______ Ensure suitable, safe command post is set up and visible.

_______ Develop and implement an incident safety plan.

_______ Perform 360-degree walkaround.

_______ Ensure appropriate use of protective clothing/equipment by all members.

_______ Ensure personnel accountability system is being utilized appropriately.

_______ Conduct rapid emergency incident risk management analysis.

- _______ Risk a lot only to save a lot.
- _______ Risk a little only to save a little.
- _______ Risk nothing to save what is already lost.

_______ Ensure appropriate safety zones are set up.

_______ Ensure rapid intervention crew is in place as soon as practical.

_______ Ensure provision for rehab has been made (rehab area established).

_______ Ensure transport unit is committed at the scene.

_______ Consider the need for additional incident safety officers.

Notes/comments or PIA issues (positive and negative):

HAZARDOUS MATERIALS INCIDENTS

_______ All members briefed on incident action plan and incident safety plan.

_______ Adequate resources at scene.

    _______ Trained personnel.

    _______ Specialized equipment.

    _______ Technical experts (OSHA, etc.).

_______ Product identified and health hazards addressed.

_______ Ensure hot and warm zones are designated and marked.

_______ Ensure decon is set up prior to entry being made.

_______ Monitor use of appropriate personal protective clothing/equipment.

_______ Ensure proper use of hazmat incident management system positions.

_______ Establish additional incident safety officers as needed for scope of incident.

_______ Brief the rapid intervention/backup crew and decon crew.

_______ Ensure that all personnel are in crews (NO FREELANCING).

_______ Ensure that stability is addressed.

    _______ Container.

    _______ Trailer.

    _______ Tank car.

    _______ Other (specify).

_______ Ensure crews are being rehabed/rotated.

_______ Watch for complacency.

_______ Extended operations concerns.

    _______ Food.

    _______ Water/drinks.

    _______ Rest/rotation of crews.

    _______ Personal hygiene.

Other issues not covered (be specific):

    ____

    ____

    ____

Benchmarks (minutes)

30 **60** 90 **120** 150 **180** 210 **240** 270 **300**

INCIDENT SAFETY OFFICER CHECKLIST

TECHNICAL RESCUE INCIDENTS

_______ Incident safety officer wearing appropriate protective clothing/equipment, including command identification vest.

_______ Accountability tag given to incident commander.

_______ Face-to-face briefing with incident commander.

_______ Understand the incident commander's incident action plan.

_______ Ensure suitable, safe command post is set up and visible.

_______ Develop and implement an incident safety plan.

_______ Perform 360-degree walkaround.

_______ Ensure appropriate use of protective clothing/equipment by all members.

_______ Ensure personnel accountability system is being utilized appropriately.

_______ Conduct rapid emergency incident risk management analysis.

- _______ Risk a lot only to save a lot.
- _______ Risk a little only to save a little.
- _______ Risk nothing to save what is already lost.

_______ Ensure appropriate safety zones are set up.

_______ Ensure rapid intervention crew is in place as soon as practical.

_______ Ensure provision for rehab has been made (rehab area established).

_______ Ensure transport unit is committed at the scene.

_______ Consider the need for additional incident safety officers.

Notes/comments or PIA issues (positive and negative):

TECHNICAL RESCUE INCIDENTS

_______ All members briefed on incident action plan and incident safety plan.
_______ Adequate resources at scene.
  _______ Trained personnel.
  _______ Specialized equipment.
  _______ Technical experts (OSHA, etc.).
_______ Ensure hot and warm zones are designated and marked.
_______ Ensure nonessential personnel removed from area.
_______ All sources of vibration removed from the area.
_______ Product (if any) identified and health hazards addressed.
_______ All utilities locked out/tagged out.
_______ Monitor use of appropriate personal protective clothing/equipment.
_______ Ensure proper use of special operations incident management system positions.
_______ Establish additional incident safety officers as needed for scope of incident.
_______ Brief the rapid intervention/backup crew.
_______ Ensure that all personnel are in crews (NO FREELANCING).
_______ Ensure that stability is addressed.
  _______ Building.
  _______ Container.
  _______ Trench.
  _______ Other (specify).
_______ Ensure crews are being rehabed/rotated.
_______ Watch for complacency.
_______ Extended operations concerns.
  _______ Food.
  _______ Water/drinks.
  _______ Rest/rotation of crews.
  _______ Personal hygiene.

Other issues not covered (be specific):

  ____

  ____

Benchmarks (minutes):

30 **60** 90 **120** 150 **180** 210 **240** 270 **300**

F

Sample Hazardous Materials Incident Site Safety/Action Plan

Adapted from Fairfax County (VA) Fire and Rescue Department

FIRE AND EMS DEPARTMENT

Hazardous Materials Response Team

Site Safety/Action Plan

Date: ______ Time: ______ Report #: ______

I. Site Assessment

Address ____

Primary Access ____

Secondary Access ____

Location type: structure roadway storm drain waterway

(circle choice) lake/pond public property private property

Terrain conditions ____

Prevailing weather: wind direction ______ speed ______
temperature ______ (F)/(C) humidity ______
dry ( ) rain ( ) sleet/snow ( )

Population density: heavy ( ) moderate ( ) light ( ) n/a ( )

II. Hazard(s) and Risk Assessment

Description of presenting conditions ____

____

____

____

(Include type of release/quantity/type of container/condition/reactions)

Primary concern ____

____

(Describe type of potential harm: thermal/mechanical/chemical)

Secondary concern ____

____

Primary exposure ____

____

____

(Include area(s) most affected/area(s) most likely to become affected)

Chemical properties that influence the hazard __________________________

____

(Attach MSDS and/or other pertinent chemical data)

Physical hazards ____

____

Ignition sources controlled: yes ( ) no ( ) ____________________________

Flammable/explosive potential: severe ( ) moderate ( ) minor ( ) n/a ( )

Potential for incident to escalate: high ( ) moderate ( ) minor ( )
(Estimation should include presenting conditions/container integrity/
chemical behavior/control mechanisms/resources/reflex times)

Overall hazard and risk assessment rating: severe ( ) moderate ( ) minor ( )

Present conditions: stable ( ) unstable ( ) undetermined ( )

Hazard and risk assessment evaluated by ______________________________

Hazard and risk assessment reviewed by ______________________________

III. Initial Response Actions

Mode of operation: offensive ( ) defensive ( ) other ( )

(Describe actions) ____

____

____

Strategy planned to support operations ________________________________

____

____

Resources needed (immediate) ______________________________

____

____

(List by priority; Resources may include foam equipment/clean-up contractor/PD/VDOT/DES/DEQ/health dept./HMS/DPW/other hazmat assist.)

Control zones (established) hot ( ) warm ( ) cold ( )

Control zone activities/access monitored by ______________________________

hot ______________________________

warm ______________________________

cold ______________________________

(See attached definition of *control zones*)

Method(s) used to establish control zones ______________________________

____

(Banner tape/signs/cones, etc.)

Performed by ______________________________ Time ______________________________

Control zone responsibility ______________________________

IV. Decontamination (contamination reduction efforts)

Type: emergency ( ) technical ( ) advanced personal ( )
radiological ( ) other ( ) ______________________________

Location ______________________________

Identified by ______________________________

Primary contaminant ______________________________

Secondary contaminant ______________________________

Decon solution selection ______________________________

PPE selection ______________________________

Method(s) used to evaluate effectiveness of decon ______________________________

____

Wash/rinse solution evaluated ( ) method ______________________________

Disposal method(s) for contaminated waste ____________________

____

Disposition of contaminated equipment/items ____________________

____

Special considerations or unique equipment needs ____________________

____

____

Personnel assigned ____________________ ____________________

____________________ ____________________

____________________ ____________________

Monitoring equipment used: yes ( ) type used: atmospheric/radiological

List monitoring instruments used (by I.D. number) ____________________

____________________ ____________________ ____________________

Decon responsibility ____________________

V. Personal Protective Measures

EPA chemical protection level for entry: A ( ) B ( ) C ( ) D ( )
(EPA chemical protection levels defined in attached document)

PPE selection for hot zone entry ____________________

____

Thermal protection: not used ( ) used ( ) type ____________________

EPA protection level for support personnel:

Protection level	Function/location
A	____________________
B	____________________
C	____________________
D	____________________

Chemical compatibility charts/reference verified/ yes ( ) no ( )

Reference used ____

____

Type of breathing apparatus in use: SCBA ( ) supplied air ( ) A.P.R. ( )

Selection of B/A based upon (type of respiratory hazard) ________________

____

Action levels for hot zone evacuation:

LEL __________ % IDLH __________ ppm

(Any time the above action level(s) are achieved, personnel operating in the hot zone will evacuate through the contamination reduction corridor and will not commence previous operations until the conditions are reevaluated and do not exceed acceptable levels.)

Other P.P.M. considerations ________________________________

____

____

P.P.M. responsibility ________________________________

VI. Isolation/Evacuation

Approximate dimensions of hot zone ________________________

Approximate dimensions of warm zone ________________________

Required isolation/evacuation distance to hot zone ________________

Reference used ____

Public evacuation: in progress ( ) planned ( ) complete ( )

Type of public protection: protect-in-place ( ) controlled relocation ( )

(If relocated, where) ________________________________

If public is relocated under controlled conditions:

( ) Identify alert method used ________________________

( ) Identify transportation medium ________________________

If protect-in-place option is used:

( ) Attach public accountability records of area(s) affected

( ) Attach information/instruction sheet given to occupants

Progress reports give to the public at __________ hrs. __________ hrs.

Method used to inform public ____

Other jurisdiction(s) affected by evacuation ______________________________

Contact person(s), name __________________________ phone _________________

____

Location of primary exposure(s) ____

____

____

Location of secondary exposure(s) ____

____

____

Degree of exposure/use numerical rating with (1) as the most severe and (3) the least: life ( ) property ( ) environmental ( )

Incident site evacuation plan ____

____

____

____

____

______________________________ Assigned to: ____________________________

Method of notification to evacuate the incident site will be (1) radio evacuation tones, (2) three short air horn blasts from apparatus on the scene, (3) voice notification over the assigned radio channel for the incident. Evacuee(s) will immediately leave the hot zone and reconvene at the designated safe refuge area.

Safe refuge area located at ____

Accountability system follow-up/time ____________________

Evacuation responsibility ____________________

VII. EMS Support

ALS unit assigned to the HMRT ____________________

HMRT entry records completed by ____________________

Baseline vitals completed: yes ( ) time: ____________________

Pre-entry criteria evaluated by ____________________

Post-entry criteria evalulated by ____________________

Treatment area located at ____________________

Rehabilitation area located at ____________________

Emergency medical treatment info attached: yes ( ) no ( ) n/a ( )

(If not attached, where it can be located ____________________)

Hospital support/contact person ____________________ phone ____________

Which hospital(s) ____________________

Patient transportation handled by ____________________

EMS support responsibility ____________________

VIII. Communications

Command post operations on channel ________ identified as ____________

Tactical operations on channel ________ other channel(s) in use ____________

Channel dedicated to HMRT entry operations ____________________

(Entry team operations section identifies hand signal communications.)

Phone number to command post ____________________ HMU ____________

Backup communication equipment located at ____________________

Communication responsibility ____________________

IX. Entry Team Operations

Tactical objectives ____________________

____

____

____

____

List actions to be performed in sequence:

1. ____
2. ____
3. ____
4. ____
5. ____
6. ____

Will be performed by: ______________________________

Backup team (who) ______________ ______________

______________ ______________

______________ ______________

(Primary duty of backup team is rescue.)

Entry team (who) ______________ ______________

______________ ______________

______________ ______________

Entry personnel briefed of tactical actions/risk/hazards/evac plan: yes ( )

Work time for entry personnel limited to ______________ minutes.

Entry personnel briefed on decon procedures: yes ( ) by whom ________

Estimated decon time for each person ______________ minutes.

Protective hoseline in place for entry personnel: yes ( ) not required ( )

Entry checklist completed by entry officer: yes ( ) by whom ________

Grounding and bonding completed: yes ( ) by whom ____________

Resistance tested at ______ ohms, by whom ______________

Resistance reevaluated at ______ ohms, time ______________

In the event entry team members experience radio communication failure, the following hand signals will be used:

hands gripping throat ____________ trouble breathing, no air
both arms above head ____________ need immediate assistance
grip partner's arm or waist ____________ leave area immediately
one hand patting chest ____________ lost radio communications
thumbs up ____________ OK. I understand

In the event radio communications are lost with the entry team, and line-of-sight contact has been broken, the backup team will be deployed to retrieve the entry team. Entry operations will not continue until communications have been reestablished.

Other considerations/concerns for entry personnel ____________

____

____

____

Entry operations responsibility ____________

X. Environmental Monitoring

Personnel assigned to monitoring ____________ ____________

____________ ____________

____________ ____________

Monitoring instruments used: ____________ ID.# ____________

____________ ID.# ____________

____________ ID.# ____________

____________ ID.# ____________

____________ ID.# ____________

____________ ID.# ____________

Other type(s) of instrumentation used ____________

____

____

(Include name of operator.)

Monitoring plan (Include procedures, time frames, and sampling locations.):

____

____

____

____

____

Diagram sampling locations:

Monitoring plan/diagram completed by ______________________

Monitoring log attached: yes ( ) no ( ) located where ______________

Note: Action levels are identified in section V, *Personal Protection Measures.*

Environmental monitoring responsibility ______________________

XI. Incident Management Organization

Incident commander ______________________

The following individuals have been assigned by the incident commander and are responsible for the operational activities of their respective sector (assignment):

Site safety ______________________
HMRT safety ______________________
Hazmat sector ______________________
Liaison ______________________
Site control ______________________
Decon sector ______________________

Entry ______________________________
Research ______________________________
EMS sector ______________________________
Documentation ______________________________
Resource sector ______________________________
Other ______________________________
Other ______________________________
Other ______________________________

(For organizational sectoring, refer to ICS chart/appendix A.)

All components of the site safety/action plan shall apply to the guidelines and parameters of the department's *Standard Operating Procedures.* Any changes(s) or revisions of SOPs, or this plan, shall be approved by the incident commander.

Evaluation of progress/operations will be reviewed periodically by the incident commander and hazmat sector officer. Report any revision(s) to the plan, operational shortfalls, and/or significant changes with the incident.

Evaluation report #1 ______________________________

____

____

Report #1 submitted by: ____________________ time: __________

Evaluation report #2 ______________________________

____

____

Report #2 submitted by: ____________________ time: __________

Evaluation report #3 ______________________________

____

____

Report #3 submitted by: ____________________ time: __________

Evaluation report #4 ______________________________

____

____

Report #4 submitted by: ____________________ time: __________

XII. Work Site Layout/Sketch

Identify the following incident components: all control zones/contamination reduction corridor/source of contamination or vehicle configuration/monitoring points (locations)/evacuation corridor/safe refuge area/medical treatment area/command post/primary exposures/significant hazards.

Wind direction ____________ Indicate reference point: (N) (S) (E) (W)

XIII. Termination/Clean-Up Operations

The incident commander has evaluated the conditions of this incident and determined that the emergency is over, and site operations are stable. The incident operations will now transition to clean-up and recovery activities.

Incident declared stable at (time): __________________ by: __________________

Guidelines for the incident commander to consider:

> All known immediate threats to public health or the environment are eliminated.
> All releases that may cause risk to human health or the environment have been contained or are in remediation.

(Excerpts from RCRA *Stabilization Guidance Document*, 7/92)

Name of contractor(s) responsible for on-site clean-up/remediation activities:

Company: ______________________________ rep: ______________________________

Company: ______________________________ rep: ______________________________

Contractor oversight responsibility:

Responsible party (person) for cost recovery: ______________________________

Address: ____

Phone: ______________________________ other: ______________________________

ERP debriefing conducted by: ______________________________ time: ________

List units involved: ____

____

____

Follow-up needed for: ____

XIV. Emergency Procedures/Guidelines

The sector officer is responsible for reporting any emergency that occurs in his/her assigned location. The sector officer shall brief the command post and the site safety officer of the event, and request any needed assistance. The command post staff will evaluate the reported emergency and initiate appropriate action. **If the emergency is of such nature to cause imminent harm to ERP, the person reporting the emergency will immediately request P.S.C.C. to activate the evacuation tones.** All site operations will stop, and response personnel will relocate to a safe area.

All on-site evacuations will require a reassessment of hazardous conditions and any associated risk to ERP. No one will reenter the site until:

> The conditions causing the evacuation have subsided or been controlled
> Control zones have been evaluated and adjusted, if necessary
> Hazard/risk analysis and site safety procedures are reassessed
> All site personnel have been briefed of current conditions and plans
> The incident commander authorizes reentry

All site workers will be made aware of evacuation signals and the locations of the safe refuge area, and medical treatment station.

The buddy system for ERP will be strictly adhered to.

All site workers should be alert for signs and indications of ignition sources, fire, explosion potential, and any hazard not previously identified. Any find shall be reported to the site safety officer and command post for appropriate action.

The department's personal accountability system will be utilized during emergency operations.

All persons operating on-site will comply with the restrictions established for the control zones. All entries and exits from the exclusion area will be recorded.

No eating, drinking, or smoking will be permitted in the exclusion area or any suspected to be contaminated.

All personnel will follow good personal hygiene practices prior to eating and/or drinking and prior to leaving the incident site.

All personnel should avoid/minimize contact with contaminants or suspected areas of contamination.

It is the responsibility of every individual who is given a task or assignment to advise his/her supervisor or sector officer when the directives are unclear or not understood. It is imperative that no assignment be attempted when the task is not understood. When procedures are complex or complicated, personnel should practice the steps prior to entering the work area.

XV. Appendix

Appendix A	ICS Organizational Chart (Hazmat)
Appendix B	Notification List
Appendix C	Planning Meeting Record
Appendix D	Regulatory Information

G

Sample Confined-Space Entry Permit for Emergency Rescue

Courtesy Frederick County (MD) Department of Fire & Rescue Services

Confined Space Entry Permit (Pre-Entry/Entry Check List)

Date and Time Issued:______________________________
Date and Time Expires:______________________________
Job Site:______________________________
Job Supervisor:______________________________
Equipment to be worked on:______________________________
Work to be performed:______________________________

Pre-Entry (See Safety Procedure)

1. Atmospheric Checks: Time ______ Oxygen ______%
   Explosive ______% LFL Toxic ______PPM
2. Source isolation (No Entry)

	N/A	Yes	No
Pumps or lines blinded,	( )	( )	( )
disconnected, or blocked	( )	( )	( )

3. Ventilation Modification:

	N/A	Yes	No
Mechanical	( )	( )	( )
Natural Ventilation Only	( )	( )	( )

4. Atmospheric check after isolation and ventilation:
   Oxygen ______% > 19.5%
   Explosive ______% LFL < 10%
   Toxic ______PPM < 10 PPM H_2S

If conditions are in compliance with the above requirements and there is no reason to believe conditions may change adversely, then proceed to the Permit Space Pre-Entry Check List. Complete and post with this permit. If conditions are not in compliance with the above requirements or there is reason to believe that conditions may change adversely, proceed to the Entry Check-List portion of this permit.

Entry (See Safety Procedure)

1. Entry, standby, and back up persons:

	Yes	No
Successfully completed required training?	( )	( )
Is it current?	( )	( )

2. Equipment:

	N/A	Yes	No
Direct reading gas monitor - tested	( )	( )	( )
Safety harnesses and lifelines for entry and standby persons	( )	( )	( )
Hoisting equipment	( )	( )	( )
Powered communications	( )	( )	( )
SCBA's for entry and standby persons	( )	( )	( )
Protective Clothing	( )	( )	( )
All electric equipment listed Class I, Division I, Group D and Non-sparking tools	( )	( )	( )

3. Rescue Procedure:
   ____
   ____
   ____

We have reviewed the work authorized by this permit and the information contained here-in. Written instructions and safety procedures have been received and are understood. Entry cannot be approved if any squares are marked in the "No" column. this permit is not valid unless all appropriate items are completed.
Permit and Check List Prepared By: (Supervisor)__________________
Approved By: (Unit Supervisor)__________________
Reviewed By (Confined Space Operations Personnel): (printed name and signature)__________________ __________________
This permit to be kept at job site.

Entry Permit

______ Confined Space ______ Hazardous Area
Permit valid for 8 hours only. All copies of permit will remain at job site until job is completed.
Site Location and Description ______
Purpose of Entry______
Supervisor(s) in charge of crews Type of Crew Phone ##

______

______

* Bold denotes minimum requirements to be completed and reviewed prior to entry *

Requirements Completed	Date	Time
Lock Out/De-energize/Try-out	____	____
Lines(s) Broken-Capped-Blanked	____	____
Purge-Flush and Vent	____	____
Ventilation	____	____
Secure Area (Post and Flag)	____	____
Breathing Apparatus	____	____
Resuscitator - Inhalator	____	____
Standby Safety Personnel	____	____
Full Body Harness w/"D" ring	____	____
Emergency Escape Retrieval Equipment	____	____
Lifelines	____	____
Fire Extinguishers	____	____
Lighting (Explosive Proof)	____	____
Protective Clothing	____	____
Respirator(s) (Air Purifying)	____	____
Burning and Welding Permit	____	____

Note: Items that do not apply enter N/A in the blank.
** Record continuous monitoring results every 2 hours

Continuous Monitoring** Test(s) to be Taken	Permissible Entry Level					
Percent of Oxygen	**19.5% to 23.5%**	____	____	____	____	____
Lower Flammable Limit	**Under 10%**	____	____	____	____	____
Carbon Monoxide	**+ 35 PPM**	____	____	____	____	____
Aromatic Hydrocarbon	+ 1 PPM * 5 PPM	____	____	____	____	____
Hydrogen Cyanide	(Skin) * 4 PPM	____	____	____	____	____
Hydrogen Sulfide	+10 PPM *15 PPM	____	____	____	____	____
Sulfur Dioxide	+ 2 PPM * 5 PPM	____	____	____	____	____
Ammonia	*35 PPM	____	____	____	____	____

* Short-term exposure limit: Employee can work in the area up to 15 minutes.
+ 8 hr. Time Weighted Average: Employee can work in area 8 hrs. (longer with appropriate respiratory protection).
Remarks:______
Gas Tester Name & Check # Instrument(s) Used Model &/or Type Serial &/or Unit #

______ ______ ______ ______

______ ______ ______ ______

Safety Standby Person Is Required for all Confined Space Work
Safety Standby Person(s) Check # Name of Safety Standby Person(s)

______ ______ ______

______ ______ ______

Supervisor Authorizing Entry______
All Above Conditions Satisfied______
Department______ Phone______

H

Public Safety Officers Benefit Program Fact Sheet

U.S. Department of Justice
Office of Justice Programs
Bureau of Justice Assistance

BJA Bureau of Justice Assistance Fact Sheet

Nancy E. Gist, Director

Public Safety Officers' Benefits Program

History

The Public Safety Officers' Benefits (PSOB) Act (42 U.S.C. 3796, et seq.) was enacted in 1976 to assist in the recruitment and retention of law enforcement officers and firefighters. Specifically, Congress was concerned that the hazards inherent in law enforcement and fire suppression and the low level of state and local death benefits might discourage qualified individuals from seeking careers in these fields, thus hampering the ability of communities to provide for public safety.

The PSOB Act was designed to offer peace of mind to men and women seeking careers in public safety and to make a strong statement about the value American society places on the contributions of those who serve their communities in potentially dangerous circumstances.

The resultant PSOB Program, which is administered by the Bureau of Justice Assistance (BJA), presents a unique opportunity for the U.S. Department of Justice; federal, state, and local public safety agencies; and national public safety organizations to become involved in

PSOB Service Standards Commitment

The mission of the PSOB staff is to assist public safety officers, their agencies, and their families before, during, and after a tragedy occurs. Three core values guide our daily operations and measure our performance. They are:

- ❑ We will respond rapidly and accurately to PSOB death and disability benefits claims.
- ❑ We will be humane in our support of public safety officers, their agencies, and their families.
- ❑ We will seek and pursue opportunities to expand our assistance to the public safety field.

To improve our response time, we continuously assess our allocation of staff and organizational processes. We will respond to the public safety field within 2 weeks once an eligible death benefits case is complete, within 4 weeks once an ineligible death benefits case is complete, and within 6 weeks once a disability case is complete. To ensure accuracy, we will use medicolegal experts and independent legal analyses from outside the PSOB Program.

To provide our services in the most sensitive and professional manner, PSOB staff receive training on key issues associated with grief, critical incident stress, and posttraumatic stress disorder. We also solicit and use information provided to us on the tone and impact of our verbal and written communication with the public safety field.

One example of the PSOB Program giving more to the field is a series of regional training sessions conducted to help law enforcement agencies prepare for the loss of an officer. It is essential that all public safety agencies be prepared to effectively assist the family, fellow officers, and the community to move forward in the aftermath of a tragedy.

Our commitment to support the public safety community has never been stronger, and it will continue to grow.

promoting the protection of public safety officers before tragedies occur. Each year, the PSOB Program receives substantial information about line of duty deaths that is used to enhance public safety officer training. The PSOB Program also encourages public safety agencies to adopt model policies that can help guide an agency through the tragic event of a line of duty death.

PSOB Program Benefits

The PSOB Program provides a one-time financial benefit to the eligible survivors of public safety officers whose deaths are the direct and proximate result of a traumatic injury sustained in the line of duty. The benefit was increased from $50,000 to $100,000 for deaths occurring on or after June 1, 1988. Since October 15, 1988, the benefit has been adjusted each year on October 1 to reflect the percentage of change in the Consumer Price Index. For fiscal year 1999, the benefit is $143,943.

The PSOB Program provides the same benefit to public safety officers who have been permanently and totally disabled by a catastrophic personal injury sustained in the line of duty if that injury permanently prevents the officer from performing ***any*** gainful work. Medical retirement for a line of duty disability does not, in and of itself, establish eligibility for PSOB benefits.

Since 1977, on average, the PSOB Program has received 275 benefit claims each year for line of duty deaths of public safety officers. PSOB Program staff respond rapidly and with sensitivity to requests for assistance from claimants and public safety agencies. They also provide moral support and, when necessary, referrals to organizations such as Concerns of Police Survivors (COPS) and the National Fallen Firefighters Foundation (NFFF), which can provide long-term support for surviving family members and coworkers of deceased public safety officers.

PSOB Program Effective Dates

The effective dates for PSOB Program benefits are as follows:

Death Benefits

- ❑ State and local law enforcement officers and firefighters are covered for line of duty deaths occurring on or after September 29, 1976.
- ❑ Federal law enforcement officers and firefighters are covered for line of duty deaths occurring on or after October 12, 1984.
- ❑ Members of federal, state, and local public rescue squads and ambulance crews are covered for line of duty deaths occurring on or after October 15, 1986.

Disability Benefits

Federal, state, and local law enforcement officers, firefighters, and members of public rescue squads and ambulance crews are covered for catastrophic personal injuries sustained on or after November 29, 1990. The public safety officer must be separated from his or her employing agency for medical reasons, and must be receiving the maximum allowable disability compensation from his or her jurisdiction, in order to initiate a claim for PSOB disability benefits. Eligible officers may include persons who are comatose, in a persistent vegetative state, or quadriplegic.

Public Safety Officers Eligible for PSOB Program Benefits

Under the PSOB Program, a *public safety officer* is a person serving a *public agency* in an official capacity, with or without compensation, as a law enforcement officer, firefighter, or member of a public rescue squad or ambulance crew. *Law enforcement officers* include, but are not limited to, police, corrections, probation, parole, and judicial officers. *Volunteer firefighters* and *members of volunteer rescue squads and ambulance crews* are covered under the program if they are officially recognized or designated members of legally organized volunteer fire departments, rescue squads, or ambulance crews.

A *public agency* is defined as the United States; any U.S. state; the District of Columbia; the Commonwealth of Puerto Rico; any U.S. territory or possession; any unit of local government; any combination of such states or units; and any department, agency, or instrumentality of the foregoing. To be eligible for benefits, a public safety officer's death or total and permanent disability must result from injuries sustained in the line of duty. *Line of duty* is defined in the PSOB regulations (28 CFR 32) as any action that the public safety officer whose primary function is crime control or reduction, enforcement of the criminal law, or suppression of fires is authorized or obligated by law, rule, regulation, or condition of employment or service to perform. Other public safety officers—whose primary function is not law enforcement or fire suppression—must be engaged in their *authorized* law enforcement, fire suppression, rescue squad, or ambulance duties when the fatal or disabling injury is sustained.

Survivors Eligible for Program Death Benefits

Once BJA approves a claim for death benefits, the benefit will be paid to eligible survivors in a lump sum, as follows:

- *If there are no surviving children of the deceased officer*, to the surviving spouse.
- *If there is a surviving child or children and a surviving spouse*, one-half to the child or to the children in equal shares and one-half to the surviving spouse.
- *If there is no surviving spouse*, to the child or in equal shares to the children.
- *If none of the above apply*, to the parent or in equal shares to the parents.

Under the PSOB Act, *child* is defined as any natural child who was born before or after the death of the public safety officer, or who is an adopted child or stepchild of the deceased public safety officer. At the time of death, the *child* must be 18 years of age or younger; or 19 through 22 years of age and pursuing a full-time course of study or training, if the child has not already completed 4 years of education beyond high school; or 19 years or older and incapable of self-support due to a physical or mental disability.

For PSOB Program benefits to be paid, a public safety officer must be survived by an eligible survivor; public safety officers cannot predesignate their beneficiaries.

PSOB Program Limitations and Exclusions

No PSOB Program benefit can be paid:

- If the death or permanent and total disability was caused by the intentional misconduct of the public safety officer or if the officer intended to bring about his or her own death or permanent and total disability.
- If the public safety officer was voluntarily intoxicated at the time of death or permanent and total disability.
- If the public safety officer was performing his or her duties in a grossly negligent manner at the time of death or permanent and total disability.
- To a claimant whose actions were a substantial contributing factor to the death of the public safety officer.
- To noncivilian members of the military serving as law enforcement officers, firefighters, or rescue squad or ambulance crew members, or to any of their survivors.

PSOB benefits do not cover death or permanent and total disability resulting from stress; strain; occupational illness; or a chronic, progressive, or congenital disease (such as heart or pulmonary disease), unless there is a traumatic injury that is a substantial contributing factor in the death or permanent and total disability. Medical proof of the traumatic injury (such as a blood test for carbon monoxide poisoning) may be necessary for coverage in these cases.

Reduction of Benefits

The PSOB Program benefit is reduced by certain payments made under the District of Columbia Code and may itself reduce benefits under Section 8191 of the federal Employees' Compensation Act. However, state and local benefits must not be reduced by benefits received under the PSOB Act, and the PSOB benefit is not reduced by any benefit received at the state or local level.

Interim Payment

If BJA determines an urgent claimant need before the final action of paying a death benefit, an interim benefit payment not exceeding $3,000 may be made to the eligible survivor(s) if it is probable that the death is compensable.

Attachment and Tax Exemption

PSOB death and disability benefits are not subject to execution or attachment by creditors. The Internal Revenue Service (IRS) has ruled that the benefit is not subject to federal income tax (IRS Ruling No. 77–235, IRS 1977–28) or to federal estate tax (IRS Ruling No. 79–397).

Attorneys' Fees

The PSOB Act authorizes BJA to establish the maximum fee that may be charged for services rendered to the claimant by another party in connection with any PSOB claim filed with BJA. Contracts for a stipulated fee and contingent fee arrangements are prohibited by PSOB regulations (28 CFR 32.22(b)). BJA assumes no responsibility for payment of claimant attorney fees (28 CFR 32.22(d)).

Filing a Claim

Eligible survivors or disability claimants may file claims directly with BJA or through the public safety agency, organization, or unit in which the public safety officer served. In most cases, the public safety agency provides BJA with sufficient information to determine whether the circumstances of the death or permanent and total disability support a benefit payment. The public safety agency prepares a Report of Public Safety Officer's Death or Permanent and Total Disability Claim Form to accompany the claim for death benefits completed by the eligible survivor(s) or, in the case of disability claims, the prerequisite disability certification package completed by the injured officer. BJA will determine whether and to whom a benefit should be paid.

For Further Information

For more information about the Public Safety Officers' Benefits Program or to share your observations and recommendations, please contact:

U.S. Department of Justice Response Center
1–800–421–6770 or 202–307–1480

Response Center staff are available Monday through Friday, 9 a.m. to 5 p.m. eastern time.

Bureau of Justice Assistance
Public Safety Officers' Benefits Program
810 Seventh Street NW.
Washington, DC 20531
202–307–0635
Toll Free: 1–888–SIGNL13 (744–6513)
Fax: 202–307–3373
World Wide Web: http://www.ojp.usdoj.gov/BJA/

Bureau of Justice Assistance Clearinghouse
P.O. Box 6000
Rockville, MD 20849–6000
1–800–688–4252
Fax: 301–519–5212
E-mail: look@ncjrs.aspensys.com

FS000066
February 1999

U.S. Department of Justice
Office of Justice Programs
Bureau of Justice Assistance

Washington, DC 20531

Official Business
Penalty for Private Use $300

PRESORTED STANDARD
POSTAGE & FEES PAID
DOJ/BJA
PERMIT NO. G–91

**Public Safety Officers' Benefits Program
Fact Sheet**

Index